PRIMARY SPACE PROJECT
RESEARCH REPORT

January 1991

Electricity

by
JONATHAN OSBORNE, PAUL BLACK,
MAUREEN SMITH and JOHN MEADOWS

LIVERPOOL UNIVERSITY PRESS

First published 1991 by
Liverpool University Press
PO Box 147, Liverpool, L69 3BX

British Library Cataloguing in Publication Data
Data are available

ISBN 0 85323 486 8

Printed and bound by
Antony Rowe Limited, Chippenham, England

CONTENTS

Introduction

This introduction is common to all SPACE topic reports and provides an overview of the project and its programme.

The Primary SPACE project was a classroom-based research project which aimed to establish

- the ideas which primary school children have in particular science concept areas;

- the possibility of children modifying their ideas as a result of relevant experiences.

The research was funded by the Nuffield Foundation and was being conducted at the the two centres, the Centre for Research in Primary Science and Technology, Department of Education, Liverpool University and the Centre for Educational Studies, King's College, London University. The joint directors were Professor Wynne Harlen and Professor Paul Black. The project had one full-time researcher, based in Liverpool, and was supported by a range of other personnel (refer to project team) . Three local education authorities were involved: Inner London Education Authority, Knowsley and Lancashire.

The project was based on the view that children develop their ideas through the experience they have. With this in mind, the Project had two main aims: firstly, to establish (through and elicitation phase) what specific ideas children have developed and what experience might have led children to hold these views; and secondly, to see whether, within a normal classroom environment, it was possible to encourage a change in the ideas in a a direction which will help children develop a more "scientific: understanding of the topic (the intervention phase) .

Eight concept areas have been studied:

Electricity
Evaporation and Condensation
Everyday changes in non-living materials
Forces and their effect on Movement
Growth
Light
Living things' sensitivity to their environment
Sound.

The project was run collaboratively between the University research teams, local education authorities and schools, with participating teachers playing an active role in the development of the project work. Over a two year life-span of the project a close relationship was established between the University researchers and the teachers, resulting in the development of the techniques which advance both classroom practice and research. These methods provided opportunities, within the classroom, for children to express their ideas and develop their thinking with the guidance of a teacher, and also help researchers towards a better understanding of children's thinking.

The Phases of the Research

Each phase, particularly the pilot work, was regarded as developmental; techniques and procedures were modified in the light of experience. The modifications involved a refinement of both exposure materials and the techniques used to elicit ideas. This flexibility allowed the project team to respond to unexpected situations and to incorporate useful developments into the programme.

There were three main aims of thepilot phase. Firstly, to trial the techniques used to establish children's ideas.; secondly, to establish the range of ideas held by primary school children; and thirdly, to familiarise the teachers with the classroom techniques being employed in the project. This third aim was very important since teachers were being asked to operate in a manner which, to many of them, was very different from their usual style. By allowing teachers a 'practice run', their initial apprehensions were reduced, and the project rationale became more familiar. In other words, teachers were being given the opportunity to incorporate the project techniques into their teaching, rather than having them imposed upon them.

In the exploration phase, children engaged with activities set up in the classroom for them to use, without any direct teaching. The activities were designed to ensure that a range of fairly common experiences (with which children might well be familiar form their everyday lives) was uniformly accessible to all children to provide a focus for their thoughts. In this way, the classroom activities were to help children articulate existing ideas rather than to provide them with novel experiences which would need to be interpreted.

Each of the topics studied raised some unique issues of technique and these distinctions led to the exploration phase receiving differential emphasis. Topics in which the central concepts involved long-term, gradual changes, e.g. 'Growth', necessitated the incorporation of a lengthy exposure period in the study. A much shorter period of exposure, directly prior to elicitation was used with 'Light' and 'Electricity', two topics involving 'instant' changes.

During the Exploration, teachers were encouraged to collect their children's ideas using informal classroom techniques. These techniques were:

i.   Using log-books (free writing/drawing)

Where the concept area involved long-term changes, it was suggested that children should make regular observations of the materials, with the frequency of these depending on the rate of change. The log-books could be pictorial or written, depending upon the age of the children involved and any entries could be supplemented by teacher comment if the children's thoughts needed explaining more fully. The main purposes of these log-books were to focus attention on the activities and to provide an informal record of the children's observations and ideas.

ii.   Structured writing/drawing

Writing or drawing produced in response to a particular question were extremely informative. This was particularly so when the teacher asked children to clarify their diagrams and themselves added explanatory notes and comments where necessary, after seeking clarification from children. Teachers were encouraged to note down any comments which emerged during the dialogue, rather than ask children to write them down themselves. It was felt that this technique would remove a pressure from children which might otherwise have inhibited the expression of their thoughts.

iii.   Completing a picture

Children were asked to add the relevant points to a picture. This technique ensured that children answered the question posed by the project team. and reduced the possible effects of competence in drawing skills on ease of expression of ideas.

iv.   Individual discussion

The structured drawing provided valuable opportunities for teachers to talk to children and to build up a picture of each child's understanding.

It was suggested that teachers use an 'open-minded' questioning style with their children. The value of listening to what children said, and of respecting their responses was emphasised, as was the importance of clarifying the meaning of words children used. This style of questioning caused some teachers to be concerned, that by accepting any response whether right or wrong, they might be implicitly reinforcing incorrect ideas. The notion of ideas being acceptable and yet provisional until tested was at the heart of the project. Where this philosophy was a novelty, some conflict was understandable.

In the elicitation phase, the project team collected structured data through individual interviews and work with small groups. The individual interviews were held with a random, stratified sample of children to establish the nature and frequencies of ideas held. The same sample of children were interviewed pre- and post-intervention so that any shifts in ideas could be identified.

The elicitation phase produced a wealth of different ideas from children, and led to some tentative insights into the experiences which could have led to the genesis of these ideas. During the intervention, teachers used this information as a starting point for classroom activities, or for interventions which were intended to lead children to extend their ideas. In schools where a significant level of teacher involvement was possible, teachers were provided with a framework to guide their structuring of activities appropriate to their class. Where opportunities for exposing teachers to project techniques were more limited, teachers were given a package of activities which had been developed by the project team.

Both the framework and the intervention activities were developed as a result of preliminary analysis of the pre-intervention elicitation data.

The intervention strategies were:

(a)   *Encouraging children to test their ideas*

It was felt that, if pupils were provided with the opportunity to test their ideas in a scientific way, they might find some of their ideas to be unsatisfying. This might encourage children to develop their thinking in a way compatible with greater scientific competence.

(b)   *Encouraging children to make more specific definitions for particular key words*

Teachers were asked to make collections of objects which exemplified particular words, thus enabling children to define words in a relevant context, through using them.

(c)   *Finding ways to make imperceptible changes perceptible*

Long-term, gradual changes in objects which could not readily be perceived were problematic for many children. Teachers endeavoured to find appropriate ways of making these changes perceptible. For example, the fact that a liquid could 'disappear' visually and yet still be sensed by the sense of smell - as in the case of perfume - might make the concept of evaporation more accessible to children.

(d)   *Testing the 'right' idea alongside children's own ideas*

Children were given activities which involved solving a problem. To complete the activity, a scientific idea had to be applied correctly, thus challenging the child's notion. This confrontation might help children to develop a more scientific idea.

In the post-intervention elicitation phase the project team collected a complementary set of data to that from the pre-intervention elicitation by re-interviewing the same sample of children. The data were analysed by to identify changes in ideas across the sample as a whole and also in individual children.

These four phases of the Project work form a coherent package which provided opportunities for children to explore and develop their scientific understanding as a part of classroom activity, and enables researchers to come nearer to establishing what conceptual development it is possible to encourage within the classroom and the most effective strategies for its encouragement.

The implications of the research

The SPACE project developed a programme which raised many issues in addition to those of identifying and changing children's ideas in a classroom context. The question of teacher and pupil involvement in such work became an important part of the project, and the acknowledgement of the complex interactions inherent in the classroom has led to findings which report changes in teacher and pupil attitudes as well as in ideas. Consequently, the central core of activity, with its pre- and post-test design, should be viewed as just one of the several kinds of change upon which the efficacy of the project must be judged.

The following pages provide a detailed account of the development of the Electricity topic, the project findings and the implications which they raised for science education.

1. Previous Research

Children's understanding of electricity and associated concepts has been an active field of research during the past decade. Most of the work reported has been conducted internationally in New Zealand, the U.K and America. The work has arisen as part of the general interest in the 'alternative conceptions' movement and has provided valuable insights into the difficulties faced by children in understanding the scientific concepts commonly presented in classrooms.

Early work was done by Andersson and Karqvsst (1979) who presented the diagrams shown in Fig 1.1 to two groups of thirty four 15 year olds and asked them whether they thought the lamp would light or not.

Fig 1.1

The results were revealing, showing firstly that despite instruction, a large number of children were unable to correctly predict which arrangement would light the bulb. Moreover, they are an important indicator of the effect of context. Faced with the MES bulb lacking any cue to indicate the presence of two terminals, large numbers of children resort to a model which sees the battery as a 'source' of electricity and the bulb as a 'sink' which need merely be connected to function. This item is particularly interesting in that an inappropriate contextual cue limits the success on this item. Even with the appropriate cue, there was a significant percentage of children who failed to provide the correct response. The result is even more remarkable in that all the children had received instruction deploying MES bulbs in their experimental work.

The work of Tiberghien and Delacôte (1976), Fredette and Lockhead (1980), Osborne (1981) and Shipstone (1984) has lead to the identification of five common models that children hold about electric circuits. These can be summarised as follows.

a. Unipolar.

In this model, the current is supplied from one terminal of the battery only and this is all used up by the device to which it is connected by a single wire. Any other wire is not considered necessary or is of no consequence. This model has been identified with an understanding which sees the battery as a source of electricity and the bulb as a sink which consumes the electricity.

b. The series or attenuation model

In this model, the child recognises that an electric circuit needs two wires to function and that the electricity circulates in one direction only. However, more current leaves one terminal than returns at the other as electricity is seen as being 'used up' by bulbs etc. In a circuit with more than one bulb, the first device uses a disproportionate share of the electricity.

c. The Sharing model

This is simply a variation of the previous model. In this version, the current is still used up by the bulbs/resistors but each one uses equal amounts of current.

d. The Clashing Currents models

Here the child explains the behaviour of the circuit in terms of two currents which leave via both terminals travelling in opposite directions. The currents meet in the bulb and mix to produce light and heat. Clearly this model has its origins in the notion that positive and negative electricity are two different 'ingredients' of electricity which must be mixed to produce any effect.

e. The Scientific model

This model sees electric charge as a means of transferring energy between one point and another. A complete circuit is required and the rate of flow of charge is the same at all points in the circuit. A full description of this model would examine the role of the battery in establishing an electric field throughout the conductor and the interaction of the electric charges with the electric field.

Both Shipstone (1984) and Osborne (1980) have conducted large scale surveys of the proportions of each model held by schoolchildren. Both results are similar although the latter's research had a larger sample size which would imply that the results are more reliable.

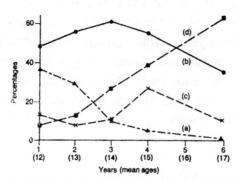

Figure 3.4: Variations in popularity of some conceptual models for current. (a) The clashing currents model, (b) all unidirectional non-conservation models, (c) the sharing model, (d) the scientific model.

Fig 1.2. Chart showing range of Models about Electric Circuits held by children from 'Electricity in Simple Circuits' by David Shipstone in CHILDREN'S IDEAS IN SCIENCE, edited by Rosalind Driver, Edith Guesne and Andree Tiberghien, Open University Press, 1985.

Not surprisingly, the scientific model which requires making a distinction between energy transfer and its means of transfer - mobile electric charge, is only held by a minority of pupils. Both studies show that the development of the scientific model is barely influenced by relatively extensive periods of instruction in electricity which occur in schools during the secondary phase of education. In addition, both report finding that some of the other models have persisted even with first year undergraduate or post-graduate teachers training to be teachers. Effectively more evidence of the strength of these alternative conceptions or the lack of an appropriate pedagogy.

Innovative approaches to Teaching Electricity

Cosgrove et al (1985) reported on a three phase teaching scheme designed to promote conceptual conflict with respect to their understanding of models of electric circuits for a group of 15 eleven year olds. This consisted of a 'familiarisation' phase, a 'challenge' phase and an 'application' phase. Their data shows that whereas only 7% of children chose the scientific model before the critical lesson, 86% chose it after. However, one year later, considerable regression had taken place as only 47% chose the scientific model to explain the behaviour of electric currents in circuits.

Another attempt to address the conceptual difficulties in understanding electric circuits was described by Shipstone and Gunstone (1985). They reported an attempt with a group of 25 twelve to thirteen year olds using an approach that was based on the assumption that most children are operating with the 'source-consumer' model. This idea is more akin to the scientist's notion of electrical energy rather than electrical current. Their programme of activities was designed to challenge this conception and to develop a discrimination between current and energy in a circuit. The results of the research were disappointing in that no significant change was produced in the understanding developed by the experimental group compared to the control group, although they did outperform the control group in most cases. The proportion of pupils showing long term retention was similar to that reported by Cosgrove et al. (1985). Commenting on the fact that no pupil was successful in more than three questions, Shipstone (1988) argued that this result was indicative of a superficial schema which lacks applicability in a wide range of circumstances and that one of the primary reasons was a lack of any holistic model of the circuit which views the system in its entirety rather than in terms of individual components and their function . Children's thinking about electrical circuits adopted sequential processing which examines the effect of each component in turn.

Haertel (1985, 1987) argued strongly that the failure of many childen to understand the behaviour of an electric circuit is due to the use of inappropriate models. The idea of particles transporting energy places an emphasis on the particles themselves, often through the use of vehicular or traffic analogies. A proper treatment of the circuit would consider it as a system where every particle is inter-related to the others. Such an approach would use the bicycle chain, conveyor belts or central heating systems as more appropriate analogies. Although his ideas were tried out in the classroom, no data were provided on the potential for such an approach to develop an improved understanding of electric circuits.

Steinberg (1987) considered that the fundamental problem for children with electric circuits is a phenomenological experience which lacks any sense of causality. The rapid rise in current in a circuit when the switch is closed, fails to provide the opportunity for observing the flow of charge in the circuit. He advocated the use of large capacitors (greater than 1F) which show transient phenomena and force children to consider the flow of charge in the circuit. In an earlier paper (1985), this approach was supported by limited data (n=18) which shows that conceptual change has occurred for the majority of the students.

Epistemological Issues

Monk (in press) has used the data collected by Shayer and Adey (1981) for the distribution of children across developmental stages to suggest that childrens' alternative conceptions can best be explained from a genetic epistomological approach. His basic assumption is that the normative development of children places inherent limits on their potential to explain the problems used by Shipstone (1984) and Osborne (1980) restricting them to concrete models which allow them to centrate on observable features. He then argues none of their data show children exceeding these limits and that thesis is a better explanatory framework of the data. Hence whilst schematic knowledge is important and the "common-sense" reasoning deployed by children limits their understanding of scientific concepts, he argues that the ultimate limit is their ability to cope with abstraction and formal operations. Some further support for this argument is possible from the data presented by Cosgrove et al (1985) which showed a regression in the number of children able to deal with the scientific model from 86% to 47%. Monk argues that the scientific model requires the schematic processing associated with early formal operations which only 30% of children reach by age 16. Whilst this is a convincing argument, it fails to address the issue of how children can develop an appropriate schematic knowledge within such a domain upto their current genetic limits. In particular, it does not consider the principal pedagogic issue raised by the large body of research reported above: that is the application of inappropriate schematic knowledge formulated from everyday experience to problems about electricity.

Rowell and Dawson (1989) argue that novice schemata are based on observables as opposed to expert schemata which are based on explanatory principles and subsume lower level novice schemata. Novice schemata can be used as the basis of inductive generalisations e.g all electrical devices require two connections to function properly. Such generalisations must confirm to reality and can be changed or even refuted by observation. The formulation of a constructive generalisation with explanatory power e.g conservation of current in a circuit, requires logico-mathematical processing and a teaching process which emphasises the change in the knowledge framework required. Hypothetical entities e.g electrons, electric charge are inferred and used as a basis for unification and explanation. They argue that the function of preliminary work in science is to prepare a common schema of sufficient complexity for the formulation of constructive generalisations.

Duit (1985) provides an example for their argument when he postulates that one of the fundamental problems is that the term 'current' is a theoretical generic idea which is inferred from observable effects such as heating or lighting in a circuit. He argues that children then perceive it as an entity which can be stored, moved or used up like other everyday entities. This is a fallacious generalisation which explains some of the explanations provided by children.

An important illustration of the effect of context was reported by Shipstone (1985). He reported two questions, shown below, examining which of the models (a - e) students deployed to reason about electric circuits. Interestingly, the use of the sequence model was much less common in the second item and this was indicative that contextual cueing was an important aspect of solving problems by children of this age. Shipstone used both of these items in a test and the data he presents show that children clearly see these items as distinctive items which bear no relationship to each other.

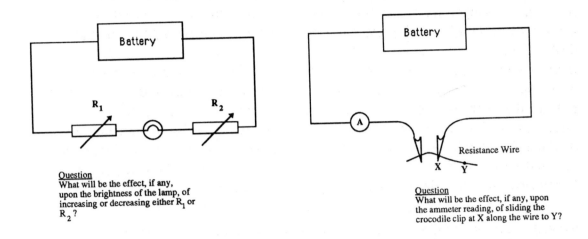

Question
What will be the effect, if any,
upon the brightness of the lamp, of
increasing or decreasing either R_1 or
R_2?

Question
What will be the effect, if any, upon
the ammeter reading, of sliding the
crocodile clip at X along the wire to Y?

Fig 1.3. Items used by Shipstone to investigate models used by children for explaining electric circuits.

Clearly the implication of this work for the research undertaken by the SPACE project was that the essential component to address with young children was to provide an experiential phase which could allow children to formulate their thinking through a variety of elicitation activities. The research aimed to chart the range and extent of the typical schematic knowledge about electricity held by young children, age 5 - 11. Further intervention activities would provide an opportunity for children to test and evaluate their models and extend and refine their schematic knowledge over a range of contexts. One of the intentions of the research was to explore what typical developments in schematic and generalised reasoning were achieved after such experiences.

References

Andersson, B. and Karrqvist, C. (1979) 'Elektriska Kretsar', EKNA Report No 2, Gothenburg University, Sweden.

Cosgrove, M., Osborne, R. and Carr, M. (1985) Children's intuitive ideas on electric current and the modificaiton of those ideas in Duit, R., Jung, W. & Von Rhoneck, C. Aspects of Understanding Electricity: Proceedings of an International Worshop. Institut für die Pädogogik der Naturwissenschaften an der Universität Kiel.

Fredette, N.H and Lockhead, J. (1980). Student conceptions of simple circuits. The Physics Teacher, 18, 194-198.

Haertel, H. (1985) The Electric Circuit as a System in Duit, R., Jung, W. & Von Rhoneck, C. Aspects of Understanding Electricity: Proceedings of an International

Worshop. Institut für die Pädogogik der Naturwissenschaften an der Universität Kiel.

Haertel, H. (1987) A Qualitative Approach to Electricity. Xerox Coporation. Palo Alto.

Jung, W. (1985) Elementary Electricity: An Epistemological look at some of the Empirical Results in Duit, R., Jung, W. & Von Rhoneck, C. Aspects of Understanding Electricity: Proceedings of an International Worshop. Institut für die Pädogogik der Naturwissenschaften an der Universität Kiel.

Monk, M. A Genetic Epistomological Approach to Research on Children's Understanding of Electricity. Private Communication.

Osborne, R.J. (1980). Children's ideas about electric current. New Zealand Science Teacher, 1981, 29, 12-19.

Rowell, J.A. and Dawson, C.J (1989) Towards an Integrated Theory and Practice for Science Teaching. Studies in Science Eduction, 16, 47 - 73.

Shipstone, D.M and Gunstone, R.F. (1985) Teaching Children to discriminate between Current and Energy in Duit, R., Jung, W. & Von Rhoneck, C. Aspects of Understanding Electricity: Proceedings of an International Worshop. Institut für die Pädogogik der Naturwissenschaften an der Universität Kiel.

Shipstone, D.M. (1984) A study of children's understanding of electricity in simple D.C.Circuits, European Journal of Science Education, 1984, 6, 185-198.

Shipstone, D.M. (1985) On children's use of conceptual models in reasoning about current electricity in Duit, R., Jung, W. & Von Rhoneck, C. Aspects of Understanding Electricity: Proceedings of an International Worshop. Institut für die Pädogogik der Naturwissenschaften an der Universität Kiel.

Shipstone, D.M. (1988). Pupil's Understanding of Simple Electrical Circuits: Some implications for instruction. Physics Education, 23, pp 92-96.

Steinberg, M. (1985). Construction of Causal Models: Experimenting with capacitor-controlled transients as a means of Promoting Conceptual Change in Duit, R., Jung, W. & Von Rhoneck, C. Aspects of Understanding Electricity: Proceedings of an International Worshop. Institut für die Pädogogik der Naturwissenschaften an der Universität Kiel.

Tiberghien, A. and Delacôte, G. (1976) Manipulations et reresentations de circuits electrique simples per ses enfants de 7 & 12 ans. Review Francais de Pedagogie, 34, 32-44.

Sample

a. Schools

Six schools from the London area were chosen for this research, all in the Inner London Education Authority. One or two teachers from each school participated in the project. Each school was allocated to one member of the research team[1] who worked closely with the teacher throughout the research phase. The research team were all equally involved on a part-time basis

The majority of the schools were selected by the science advisory teacher, Maureen Smith, who had already been working in the locality providing support to primary schools in the development of primary science work.

b. Teachers

Most of the teachers invited to participate in the project were those known to the researchers from the previous work[2] in the area of light with two exceptions where a staffing change had occurred. This was advantageous in providing a pre-existing relationship and link between researcher and teachers which could be developed. Teachers were able to use this relationship to express their uncertainties about the work and ask for clarification. Unfortunately, the local authority was unable to release any of the teachers due to the difficulties experienced during this phase in obtaining any supply cover in the London area. This meant that all meetings had to take place during the teachers' own time after school, and this had the effect of curtailing the extent of the teacher contribution to the research on this topic.

The teacher's normal style of working varied, between individuals who made sole use of classrooms organised around groups using a topic approach and an 'integrated' day, and those who preferred to keep the class working together on a common theme. Teachers were encouraged to integrate the activities into their existing mode of working as there was a limitation to the amount of change of teaching style that could be expected.

Many of the difficulties experienced and expressed by teachers with teaching science are associated with a lack of confidence in their own understanding of the background knowledge.This was particularly noticeable feature with this topic. Many of the teachers concerns could have been seen as a) a lack of a clearly internalised

---

[1]   The research reported here was undertaken by the authors, Maureen Smith, John Meadows and Jonathan Osborne on a part-time basis during 1988.

[2]   SPACE Research Report: Light. Osborne, J., Black, P.J., Smith, M. & Meadows J.M.. Liverpool University Press. 1990

model of what constituted an appropriate understanding of the behaviour of an electric circuit, and b) uncertainty about the level of understanding that it would be reasonable to expect a child to achieve. Whilst teachers understood that the research project was attempting to provide some insight into the latter question, it was clear that the degree of uncertainty was a source of anxiety for teachers.

Names of the participating teachers, their schools and head teachers are provided in Appendix 1.

c. Children

Despite the limitation to a particular locale, the schools used reflect the wide variation seen in the London area between schools based in deprived areas and those with a substantial middle-class catchment area. The children used in the sample represent children with a wide range of ability and ethnic background. All children in the classes of the participating teachers who were involved in the project to some extent were used for the pre- and post-intervention elicitation activities. Inevitably there were some children who were not present for both phases of the activity and the data collected from these children has not been used.

For the purpose of analysis, the children have been grouped by age into infants (5-7), lower juniors (8-9) and upper juniors (10-11). In case of any doubt surrounding the particular grouping of a child, the year of schooling was used to decide the appropriate cohort for a child. The size of the infant cohort was limited by the necessity to interview and transcribe all the data for these children. Some of the data from lower and upper junior children was obtained through written responses.

iv) Liaison

During this phase of the project, the research was conducted by a team of three working part-time with the schools and the relevant teachers. Each member of the team was allocated a particular school. The team would meet on a regular basis to plan and co-ordinate the research, exchange information and develop materials. One member of the team was more closely involved in the schools through her work as a science advisory teacher and was able to use this role to provide enhanced support and guidance to the teachers involved in the project. .

The Research Programme

Classroom work on the topic of 'electricity' took place over a relatively long period in the school year which can be summarised as follows.

Pilot Exploration	March 88
Pre-Intervention Data Collection	April 88

Intervention   May - June 88

Post-Intervention Data Collection   July 88

The pilot exploration phase was based on interviews with a small number of children (25). These interviews used a wide range of questions to explore the nature of children's understanding of the topic of electricity and associated concepts. In addition, drawings and answers to written questions were employed to examine how valuable and reliable such sources were for eliciting children's meanings and understanding. Sample questions are shown in Appendix 2. The exploratory nature of this phase was required to supplement what little literature there was available on the nature of young (5-11) children's understanding of this topic. Many of the tools devised for probing children's ideas were modifications of methods that had been used with older children. At the end of this phase, the data was examined to determine which were the most valuable lines of approach for eliciting children's ideas about this topic. The other valuable feature of this phase was that it provided time for developing a relationship with the teachers and the children so that they could become accustomed to the mode of working required.

Essentially, the classroom elicitation techniques were refined by the pilot phase and the experience provided an opportunity for teachers and researchers to develop familiarity with the material and each other. Data on children's ideas were then collected in classrooms using the selected activities. These questions and activities are shown in Appendix 3. The main methods of elicitation relied on written answers and children's drawings. The data were also supplemented by interviews with a few children to provide further insight. Data from infants were collected by interview and drawings as these children found it very difficult to provide written answers to questions. No attempt was made to collect interview data systematically from upper and lower juniors. This limitation was imposed by the restricted time available from the part-time researchers.

The intervention activities were designed in consultation with the teachers and from an examination of the data collected previously. The data suggested three areas of interest for possible conceptual development and a framework of activities was designed which could be used by children to test their own ideas on the behaviour of electricity. This was not presented as a prescriptive framework, but simply as a range of exercises which could be used in the classroom. Teachers and children were free to try other lines of investigation they wished to pursue. After the completion of the intervention phase, another set of elicitations was used with the children based on similar questions to those used in the elicitation prior to the intervention.

Defining 'Electricity'

Any attempt to develop a child's concepts needs to be based on a map of what a preferred understanding would be. The following list was compiled by the team to provide a map of ideas considered an a priori necessity for the development of the scientist's view.

1.   Electricity can move or flow.

2.   Electricity is required for a wide range of devices i.e heating and moving objects, providing light and making magnets.

3.   Electrical devices require two connections with wire to a battery to function.

4.   The two connections provide a complete path around which electricity can flow.

5.   Some materials allow electricity to pass through them and other materials do not. Those which do allow electricity to flow through are called conductors.

6.   The strength of electricity is dependent on the number of batteries and the voltage they supply.

7.   Electricity can be made with dynamos.

This list represents a basis or platform for the fuller understanding of the scientist. It suffers from the use of the generic term 'electricity' for electrical energy and electrical charge but such a distinction can not be made with children of the primary age range. The purpose of this list is to provide a framework or point of reference for the research. These statements represent a collection of ideas that children may develop by age 11. For example, children need to develop an understanding that two connections are required before they can understand the scientist's picture of current conservation which is generally developed in secondary schools. One of the aims of the research was to examine to what extent, as a consequence of the experiences that were provided by this research programme, such ideas could develop in children and at what ages.

These ideas also provide a framework for examining children's ideas allowing three questions to be addressed.

a)   How different were the conceptions held by many children from such a framework and how disparate are their ideas?

b)   What development was observable in children's ideas across the age range?

c)   What potential did the planned intervention have for the development of children's ideas towards the scientist's view?

This list was also used as a reference point for the development of the intervention. Given such a framework of ideas, the task was to develop activities which would assist the formation of a fuller understanding in children. The activities were devised using simple materials familiar to children. Their primary role was to provide a focus for discussion of children's thinking and to challenge their existing ideas. Other considerations in designing the activities were that the materials should be simple, easy to manipulate and safe to handle.

3. Pre-Intervention Elicitation Work

The Pilot Phase

This phase of work was conducted by the research group. Ideally it would have been preferable to train the teachers involved to perform this work. However, the lack of possible release provided little opportunity to do such training and it was judged unwise to place too many demands on teachers. Consequently, the available time with teachers was used for preparing teachers for the main intervention work. The collection of the data for the pilot phase and the elicitations were done by the research group.

The research began with an initial pilot phase during which a wide variety of experiences and questions were used with 25 children of different ages in interview situations. The purpose of this was a limited empirical study of the range and nature of responses provided by children to explain phenomena associated with electricity. The following activities were devised or used with children during this phase.

a.   *Writing three sentences about electricity*

This activity was used as an open-ended activity to explore what associations the word 'electricity' had for children. Children were asked to 'write three things about electricity'. Most children managed to write two or three sentences or features of electricity. The predominant feature of their writing was the use of electricity for a wide variety of functions. Commonly sentences would say,

'electricity is used for homes';
'electricity gives us light.';
'a heater uses electricity.';

other notable features were such sentences as

'electricity can give you a shock.';
'electricity cost a lot of money';
'electricity comes from the sun.'.

The predominant expression was the association of electricity with the functioning of a piece of machinery or light and a wide variety of uses were mentioned. This activity was conducted with infants verbally. Their responses also showed an awareness of the uses of electricity.

b. *Objects that use electricity*

The purpose of this activity was to gain some insight into the range of objects that children saw as requiring electricity to function. Children were asked to draw as many objects as they could think of that needed electricity to work, or alternatively write down the names of these object. It was hoped that this would provide some insight to the range of children's experience of objects associated with electricity and the origin of this experience. Children were able to draw a wide variety of devices which needed electricity to function. These were predominantly domestic in origin.

c. *Where does electricity come from?*

The purpose of this question was to explore any ideas children had about the origins of electricity and the models that they were using to express their ideas. Responses provided were wide ranging. Some children were able to say that electricity came from power stations and was brought on wires. When challenged that there were no visible wires in London, they suggested that they were underground. However other sources of electricity mentioned were lightning, solar plates, wind power, radioactivity, water generators and people as 'everybody has electricity in them'. However, most of the answers lacked clarity reflecting linguistic/conceptual confusion between form, function. For instance, the boxes inside/outside houses were often called 'distributors' and viewed as sources of electricity. Electricity came from transformers, sockets, wire and meters and no further explanation was provided.

d. *How is electricity made?*

The most common answer to this question was that electricity was made in power stations. However, relatively few children were able to provide this response and a range of answers associating the production of electricity with lightning, transformers and pipes were produced. One or two children had more extensive knowledge saying that it was made out of coal or that the 'black carbon thing' in the battery makes electricity.

f. *What is the difference between electricity from plugs and batteries?*

Children were generally expressed the view that electricity from plugs was stronger than that from batteries and that it would not last for ever whereas electricity from plugs did. Common responses would be to say that 'batteries run out' and that 'a plug is more powerful than electricity.'

g.   *How does a switch work?*

The purpose of this activity was to explore whether children held any model of electricity which could explain the functioning of a switch. This question posed much more of a problem for children and many were unable to give any answer. Those that did attempt an explanation would attempt to explain in terms of wires and gaps. The wires went through the wall and the switch was 'a gap'. Pressing the switch let the electricity 'run up' and turning it off made it 'run back'. The answers to this question strongly suggested that most children lacked any picture of electricity flowing other than a source-sink model where the switch acted as a gap which stopped the electricity moving.

h.   *How fast does electricity travel*

Children's general impression was that electricity travelled very fast using such expressions as 'faster than light' or 'two hundred miles an hour' to convey a general notion of 'very fast'. Few children had any difficulty in responding to this question.

e.   *Lighting a bulb*

Children were provided with batteries, wire which had been bared at the ends and a small torch bulb. They were then asked to make the bulb light and record the method that they used. This is not an easy manipulative task so children were asked to work in pairs for this activity. In order to find out whether it was the nature of the connections to the lamp bulb that were problematic for the children, the activity was repeated with a small electric motor where the two terminals were clearly defined. There is clear observational evidence that many children do not perceive the two connections of a bulb because the separation of the terminals is not clear. In addition, this activity was also repeated with the components of a Unilab junior electricity kit to examine whether presenting the problem in a different context affected the children's performance on this task. Most children had severe difficulty with this task though it posed less difficulty when using the kit materials. The strategy adopted generally tended to be a 'trial and error' strategy beginning with a single wire from battery to bulb which showed that a simple source to sink model was the starting point for most children even of a very young age. The failure of this model taxed most children and only older and more determined children were able to overcome these difficulties.

g.   *Materials that conduct electricity*

Children were shown a range of materials e.g rubber, paper clip, wax, covered and bare wire, plastic comb and asked whether they thought electricity would be able to go through them. The predominant feature of their responses was a lack of any clear idea of which materials would conduct. No child was able to predict all correctly and the idea that only metals would conduct electricity was not expressed. This would suggest that the abstract notion of a metal was not something which these children possessed.

h.   *What would be the effect of using larger, more batteries.*

This question was asked to explore whether children had any understanding of the notion of voltage. Children were shown a battery lighting a bulb and then asked what would happen if it was replaced by a larger battery. The predominant response was that it would be brighter. When shown the effect, one or two children were able to explain that the voltage was the same and therefore it made no difference. Most children stated that the larger one would last longer. The use of two batteries in series led to the statement that the lamp would be brighter because 'two batteries are stronger than one' which was a reasonable common-sense explanation of this effect.

i.   *Static phenomena.*

Several children had mentioned static effects when asked where electricity comes from earlier. Children were shown a comb being rubbed through the interviewer's hair and then being used to pick up small pieces of paper. This activity was used to explore whether children had any deeper knowledge of static electricity and its effects. Only a few children had seen the effect before and none of them attributed it to being an electrical effect. Children's experience of static electricity seemed to be limited to sparks seen on nylon clothes in the dark and shocks obtained from objects such as carpets.

Summary:

The main aim of this phase of the activity was to trial a wide variety of activities and questions which could be used for the elicitation phase. Any activities which were found to be non-productive in eliciting children's thinking on the topic were discarded for the elicitation phase. Only one activity was found to be of little value which was the attempt to explore children's knowledge of static electrical effects with the comb and pieces of paper. Few children had observed this effect and no children were found who could provide any explanation of this phenomena. Consequently, it was decided to avoid using elicitation activities which addressed electrical phenomena associated with static electricity.

4. Children's Ideas about Electricity

This chapter presents a qualitative picture of young children's thinking about electricity as found in the elicitation phase. The elicitation was carried out by teachers and researchers using a subset of the activities employed in the pilot phase with individual or small groups of children. Details of these activities are shown in Appendix 3.

The elicitation activities were a mixture of activities requiring children to provide written responses and drawings to indicate their thinking. It would have been preferable to conduct clinical interviews with all children but limited resources made this impossible. Interviews were used with infant children who were incapable of providing written responses to much of the material.

Data were obtained in six areas which can be categorised as

    a.   The properties of electricity

    b.   The uses of electricity

    c.   Making circuits

    d.   Materials that conduct electricity

    e.   Testing for conductors and insulators

    f.   Using more batteries

This section provides an overview of the main features of children's thinking in these areas.

The properties of electricity.

Partly to obtain an insight into children's thinking and partly to introduce them to the topic, children were asked to indicate general ideas about electricity by writing three sentences containing the word 'electricity'. This activity was not used with infant children who would have found it difficult to provide a written response.

In addition, a number of specific questions were asked about electricity which asked what children thought electricity was like; where it came from; how fast it travelled; how switches worked and what the distinction was between electricity from plugs and that from batteries?

The full quantitative results for their responses are presented in Chapter 6 and material discussed here is a qualitative report designed to convey the nature and style of childrens' responses.

The predominant feature of children's writing about electricity was an association of electricity with function. Typically sentences like those following were written in response to question 2 or 5 (Appendix 3).

> *'You will find if you have an electric cooker that it uses electricity. I have electricity in all of my lights'.*

> Anne: Age 10

> *'Electricity helps us in the home.'*

> Jane: Age 11.

> *Electricity is a very strong form of power, it runs all sorts of things.....it would be hard to live without it.'*

> Harry: Age 10

> *'Without electricity we would not be able to read.'*

> Solomon: Age 10

The picture that emerges from these statements is that electricity is seen as a pervasive and universal 'substance' which is required to work or power most machinery. In a stronger form, electricity is viewed as an essential prerequisite for life which is reflected in the following statements.

> *'Electricity is part of our lives'*

> Jane: Age 9.

> *'In every house, there is electricity.'*

> Kelly Anne:Age 9.

> *'Electricity is very useful. Electricity is used every day.'*

> Joseph: Age 10.

> *'We could not live without electricity'*

> Daniel: Age 10.

Such statements were more common with older children which gives some indication that these children are prepared to recognise a concept of electricity which is not associated with the functioning of specific machinery and that they were beginning to recognise electricity as an independent entity.

Another aspect of electricity that was commonly recognised was the danger. The following are representative statements of the responses provided by children.

> *'Electricity is dangerous you can kill yourself.'*
>
> Matilda: Age 6.

> *'Electricity is dangerous.'*
>
> Daniel: Age 10.

> *'Electricity can give you a shock.'*
>
> Natasha: Age 9.

> *'You could get an electric shock from electricity.'*
>
> Makeda: Age 10

This feature of children's knowledge of electricity has been documented before by Solomon, Black et al (1985)[1]. However, this was not the predominant feature of their responses and only a minority mentioned the aspect of danger associated with electricity.

Asking children what electricity was like (Q6) and where it came from (Q1) produced a wide range of responses that provided some insights into their thinking and the origins of their ideas about electricity. Many associated electricity with gas and heat. For instance in responses to Q6:

> *'If electricity has been left on for a long time, it would be very hot.'*
>
> Laurence: Age 9

> *'It makes your house warm.'*
>
> Layi: Age 7

> *'Electricity gives us warmth.'*
>
> Daniel: Age 9

The association with heat is possibly not surprising in view of the widespread use of electricity for heating. Many children's experience is of objects worked by electricity which get warm as evidenced by these statements. Other children attempted to relate electricity to gas.

> *'Electricity is hot.....fire...comes from big gas things.'*
>
> Steven: Age 7.

---

1.   Solomon , J., Black, P., Oldham, V. & Stuart H. The Pupil's view of electricity. *European Journal of Science Education,* 7,3, 281-294.

'Electricity comes from gas.'

Wayne: Age 10

'Electricity is like gas..you can't see it, it is dangerous and it helps things work.'

Wayne: Age 8

The previous statement shows an attempt to define electricity by analogy with gas as well as in terms of its effects. The association with 'gas' may also be an attempt by children to provide a more substantive concrete reference for electricity.

Several statements were collected associating electricity with burning which also helps to develop the idea that electricity is 'hot'.

'One day I was putting my light on and.....I turned to turn my light off and it burns my house. It burns...my tele was burnt.'

Layi: Age 7

'Burn you....when I was a little baby, I went to hospital.'

Danny: Age 5

'If you put a plug .....in the socket and you put it in there a million times, then it might blow and raise a fire.'

Alex: Age 6.

More statements associating electricity with fire and warmth were obtained from infant children and reflect an awareness of the danger of electricity which has probably been instilled by their parents. Interestingly, older children tended to give answers that identified some of the properties of electricity e.g

'You cannot see electricity.'

Mark: Age 9

'Electricity is like magic.'

Acima: Age 10

Children were also asked 'Where does electricity come from?' (Q1) and 'How does it get here?' (Q13) and these two questions produced a wide variety of responses. Some children responded that it came from 'power stations' or 'electricity stations'. However a considerable number associated the origin of electricity with the sun or lightning or even in the occasional case, satellites.

'Electricity is like lightning that comes from space - it hits the wires that are on the street and it goes to the top of your house and makes the telephone work. All the electricity goes down to the control box in your house.'

Farrukh: Age 8.

'Electricity is in the sun.'

Westley: Age 10.

'I think electricity gets here by satellite.'

Kelly Ann: Age 9.

Many children said that electricity got here by wires, cables or pipes or 'from underground' which presumably led to the association by a few children of electricity with water which were both seen as coming 'through pipes'.

Other responses show how younger children are attempting to make sense of the varied sources of information and observations to which they are exposed e.g

Sonia: Age 8   *'Electricity comes from God.'*

Interviewer: *'How does it get here?'*

Sonia: *'God brings it and puts it in those big round things (points to nearby gasometers).*

Interviewer: *'How does he do that without us seeing?'*

Sonia: *'He made the round things before he made people and he put electricity in them.'*

Interviewer.   *'Where does electricity come from?'*

Alex: Age 6.   *'You buy it.'*

Interviewer: *'Where from?'*

Alex: *'Shops'*

Interviewer: *'How do you get it home?'*

Alex: *'You take it home.'*

Interviewer:	*'Where does electricity come from?'*
Shantelle: Age 6.	*'It comes from a kind of house.'*
Interviewer:	*'What kind of house?'*
Shantelle:	*'All electricity in it.'*

Question 14 that was designed to explore whether children were aware of any difference between electricity from the mains and electricity from batteries generally failed to elicit any significant response from infant children other than 'Don't know'. There was some doubt as to whether infant children even perceive batteries as being associated with electricity.

Children were asked 'How fast does electricity go?' (Q9) and the predominant response to this question indicated that most children had the impression that electricity travelled very fast. Typical answers state that it went 'very, very fast' or attempted to quantify it's speed in terms of a number that was considered very fast e.g. '200 miles per hour', '30,000 miles a second.'. The occasional response indicated the reasoning underpinning this belief.

'It must go very fast....faster than Concorde because you can phone to France in about 10 seconds, so electricity can get to France that quickly.'

Robert: Age 10.

Interestingly, what this reflects is an impulse model of electricity which sees electrical phenomena as being transmitted in pulses down wires. It was hoped that the range of questions used would provide more information about children's models of electricity but the items used failed to reveal their models in greater depth. Question 10 about switches and how they functioned generally elicited disappointing answers which described switches working when pressed or 'by electricity'. Partly this was due to the question which failed to place any emphasis on the internal working of the switch, but it also revealed that very few children had any idea of what was inside a switch and how it operated. Some responses used metaphors that were consistent with a 'water model'.

'When you turn on the switch, you let electricity through.'

Daniel: Age 9

The intervention did seem to improve the knowledge of switches for some of the upper juniors. The following are typical answers to explain switches obtained after the intervention from a minority of pupils.

'When it is on, it allows electricity to pass through but when it is off, it breaks the circuit.'

Sarah: Age 10

*'When you push the switch, two wires connect to each other and one of the
wires goes to the bulb and the other goes to the cable.'*

Mark: Age 10

However such explanations were rare and most children found this question difficult
proffering a mechanistic explanation of a switch as something which worked when
pressed and many children failed to offer any response at all.

The overall picture that emerges from these responses is one in which electricity is
viewed by many children as a quasi-mystical substance whose effects can be
observed. The principal effect of electricity was an association with heating.
Electricity is also seen as being in lamps and televisions and used by cookers and
computers. It can be seen as something which could be described as a vitalistic to
machinery, essential for their functioning. The language used generally reflects a
source to sink model in which electricity is something semi-concrete which travels
fast down wires, pipes and cables.

Uses of Electricity

In both phases of the elicitation children were asked what electricity was used for.
The primary reason for asking this question was to see what items or purposes
children associated with electricity. Children mentioned a large number of uses for
electricity nearly all of which are associated with domestic uses of electricity. Tables
1, 2 and 3 show the uses mentioned by infants (5-7), lower Juniors (7-9) and upper
Juniors (9-11) respectively.

In all cases there were a large number of other items mentioned by less than 6% of
the sample. The results showed clearly the predominance of domestic items and a
remarkable consistency across the age range. Older children mentioned more uses
than younger children which may reflect a greater range of experience or
alternatively the capacity to provide more extensive answers.

In some cases, the use of electricity was associated with the general capacity to do
work with statements such as:

'We use electricity for working things.'

Mark:Age 9.

'To make big machines work.'

Joseph Age 10.

Infant Responses showing Uses of Electricity

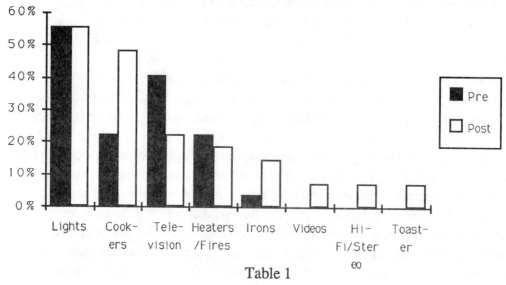

Table 1

Lower Junior Responses Showing Uses of Electricity

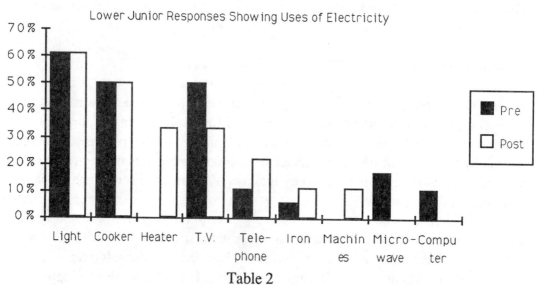

Table 2

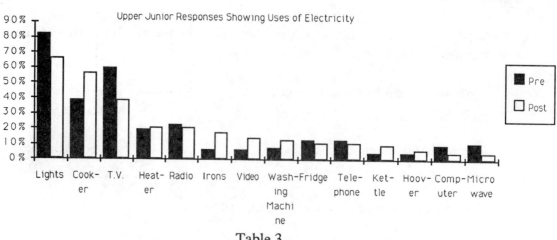

Upper Junior Responses Showing Uses of Electricity

Table 3

However, the majority of children mentioned specific uses and it was predominantly older children who offered generalised statements about the function of electricity. The range of responses showed that even young children were aware of the significance of electricity in their lives and of the wide range of uses.

Making Circuits

One of the *a priori* ideas about electricity taught by science education is the concept that electricity needs a complete circuit to flow around and without this electrical items will not function. Therefore many of the elicitation activities were designed to examine the knowledge that children had about simple electrical circuits and the models that they were employing. Their ideas were explored by providing exploratory activities with simple electrical materials i.e light bulbs, wires and batteries and then asking the children to draw the connections which would be needed to light a bulb on a pre-drawn diagram of a bulb and battery, a motor and a battery and to show if they knew how the bulb could be lit with only one wire (Q3, 4, 8). Two elicitation items provided a means of evaluating if the response of the children was consistent or context-dependent.

Children's answers fell into the following categories.

a.   A single connection.

Many children provided a drawing indicating a single connection between the battery and the bulb to show how to light the bulb. This source to sink model was produced extensively and reflects an understanding which sees the battery as a source of power, the light/ motor as the consumer and the wire as the necessary link to enable the supply. Some children drew this response even when they were aware that it failed in practice to light the bulb.

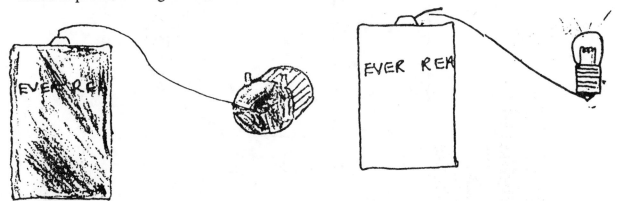

Fig 1: Tom - Age 7   Fig 2: Dano - Age 9

b.   Two battery connections, 1 device connection.

These children showed an awareness of the need for two wires coming from the the battery but were not aware of the need to join the wires to separate points on the bulb or the motor. Typical examples are shown in Fig 3 and Fig 4.

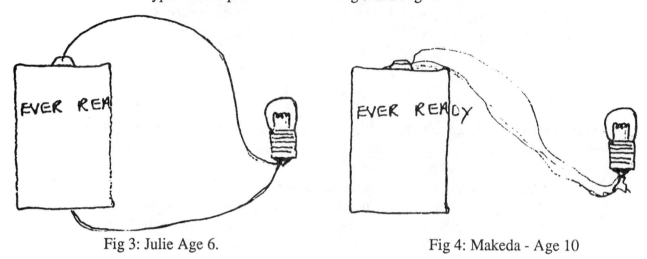

Fig 3: Julie Age 6.   Fig 4: Makeda - Age 10

Such responses were considered indicative of a more sophisticated idea about the physical requirements necessary for a circuit. Previous research has argued that such a model is consistent with the idea that electricity consists of two ingredients, positive and negative, and that children see the mixing of these two ingredients as necessary for anything to work. Such drawings would be consistent with such an idea or, more simply, they may show a failure to recognise the two connecting points on a MES bulb.

c.   *2 Battery connections, 2 device connections.*

A third type of response showed two battery connections and two device connections but in the wrong places.

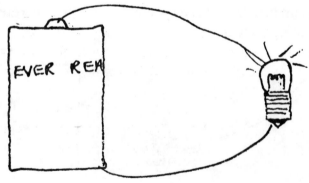

Fig 5: Hayley - Age 11.

Such responses were relatively rare and were presumed to indicate an awareness of the need to have two wires attached to different points on the device. However, there was a lack of knowledge about which points on the device the wires should be connected to.

In the case of the motor, there was a large number of responses of the type shown in Fig 6. These responses were taken to be an attempt by the child to show the correct method of completing the circuit.

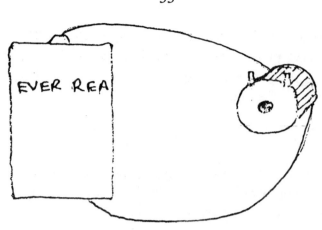

Fig 6: Zufie - Age 11

d.   Two correct connections shown

Many children were able to indicate correctly the connections necessary to make the lamp or bulb function, particularly during the post elicitation. Fig 7 and Fig 8 show typical drawings produced by such children.

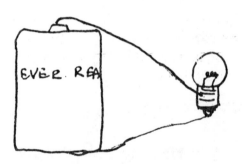

Fig 7: David - Age 5   Fig 8: Alexander - Age 6

What was notable was the change over the intervention from the predominance of unipolar models of electric circuits to models which showed a recognition of the need for a complete circuit and two wires.

One or two children indicated that they saw the operation of the circuit in terms of a flow by adding arrows to the diagrams.

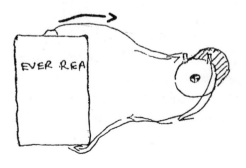

Fig 9: Harry - Age 10

However, such responses were rare and unfortunately there was insufficient time to explore what models children had of the behaviour of the electricity in the circuit and this is a key area that needs to be addressed by future research.

e.   *No response*

There were a number of children who simply failed to draw an answer to the question (Q3). No attempt was made to explore why they were unable to provide any answer but the number doing so reduced after the intervention.

Another notable aspect of children's responses was the lack consistency about their responses. A sizeable minority of infants and upper juniors and a majority of juniors, who could show successfully how to connect a motor to a battery, could not repeat this when presented with a bulb and battery or vice versa. Fig 10 shows such a response.

Fig 10: An example of inconsistent responses to similar questions asking
how the bulb/motor should be connected to work

This result is interesting in that it shows clear evidence that even within a confined domain, children's responses are dependent on context. Such behaviour has already been noted in the work undertaken previously on light[1] . These instances show that the child perceived them as being distinct, centrating on the observable concrete distinctions and lacked any model which would allow them to recognise the similarity.

Materials that conduct Electricity

One basic aspect of scientific knowledge about electricity is that some materials will conduct electricity whereas others will not. To explore the extent of children's knowledge about the ability of materials to pass or not pass electricity, a range of common materials were presented to children and the child asked whether they

---

1.   SPACE Research Report: Light. Osborne, J.F, Black, P.J., Meadows J.M & Smith, M.
        Liverpool University Press. 1990

Table 4 (a)
Showing pre and post responses for Infants for whether
materials will conduct electricity (n = 27)

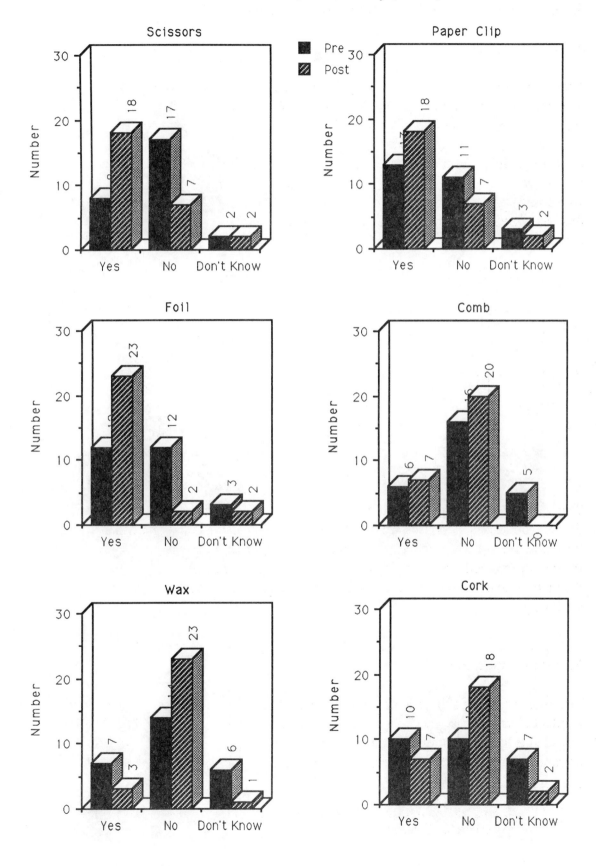

Table 4 (a)
Showing pre and post responses for Lower Juniors for whether
materials will conduct electricity (n = 18)

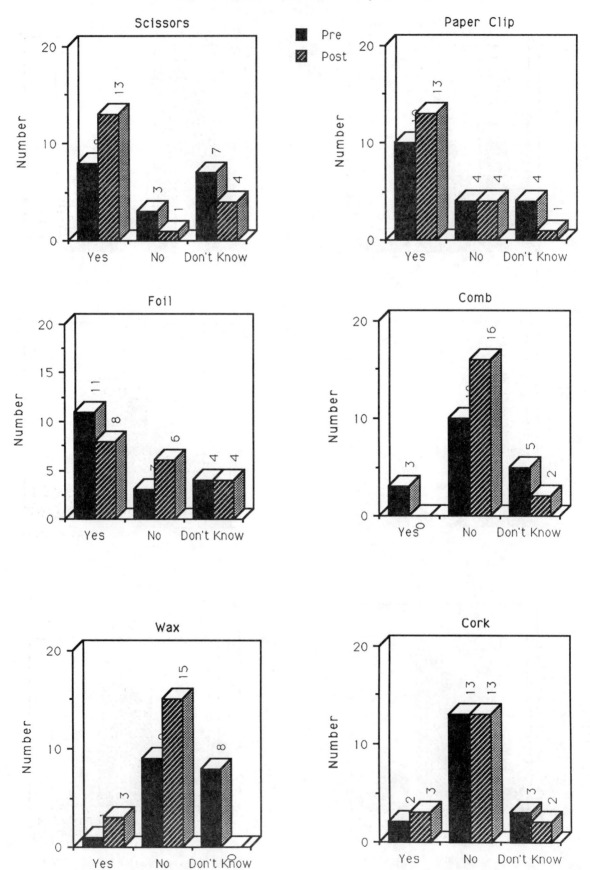

Table 4 (c) .
Showing pre and post responses for Upper Juniors for whether
materials will conduct electricity (n=62)

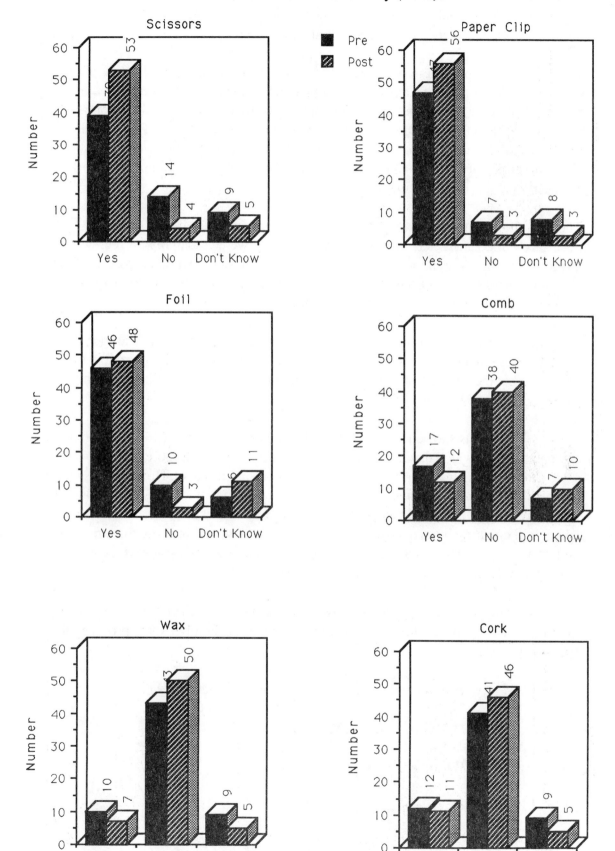

thought the material would 'let electricity through it', 'not let electricity through it' or whether they 'didn't know'. The results obtained from children are presented in tables 4(a) -(c)and discussed in more detail in Chapter 6. The data only shows responses indicated and in many cases children chose to give no response when questioned whether a specific material would pass electricity.

The six materials used were three non-conductors, a wax candle, a cork and a plastic comb and three conductors, a paper clip, a piece of kitchen foil and some household scissors. Table 4(a) shows the responses obtained from infants, table 4(b) from lower juniors and table 4(c) fromupper juniors for the materials used; Each graph shows the the number of responses obtained before and after the intervention in the three categories of 'yes-it will conduct/let electricity pass', 'no it will not conduct/let electricity pass' and 'don't know'.

The data show quite clearly that upper junior children had a clear idea of which materials will conduct electricity and that those ideas were essentially correct with a large number of children making the correct predictions about whether materials will or will not pass electricity. There was some evidence from this that a minority of children were less certain about non-conductors.

The data for lower juniors show a similar pattern though with a smaller sample, the evidence is not quite as distinct. However, even from this sample, it is possible to conclude that the majority of lower junior children were capable of distinguishing non-conductors of electricity from conductors.

The data for infants showed little evidence that children prior to the intervention had any clear idea of which materials would conduct electricity with more children saying that scissors would not conduct than those saying it would. However, it is notable that the intervention has had the effect of changing children's perceptions so that the majority of children were capable of correctly identifying those materials which will conduct electricity afterwards. These results would indicate that an understanding of which materials conduct electricity was evolving across the age range possibly as a consequence of general experience.

There was only a limited opportunity to explore with some of the infant children why their ideas had changed. Most children were unable to explain but some provided the following reasoning.

Interviewer: *'How do you know which things will let electricity pass?'*

Billy: Age 6   *'Cos you see the bulb light up.'*

Interviewer   *'Why does that happen?'*

Billy   *'Cos it's metal.'*

Interviewer: *'How do you know which things let electricity pass?'*

Danny: Age 5   *'They have all got metal'*

These excerpts show that it is possible for young children to develop the concept of metals and that one of the attributes of a metal is its ability to conduct electricity.

Testing Materials for conduction

Children were asked how they would test to see if an object would let electricity pass through it. This was done partly to see if they knew that a circuit was required and partly to test if they could represent the circuit that was needed. Children were encouraged to draw or write a response. This proved to be a difficult exercise for most children and consequently was not used with infant children who have substantial difficulty in accurate, presentational drawing let alone writing. Only upper juniors were really capable of this task and Fig 11 shows an example of such a response.

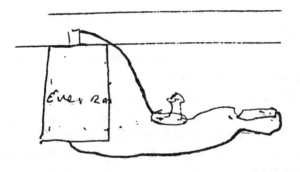

Fig 11: Simone Age 11

Many of the upper juniors used the 'circuit concept' to attempt to explain how to do this task (Fig 12).

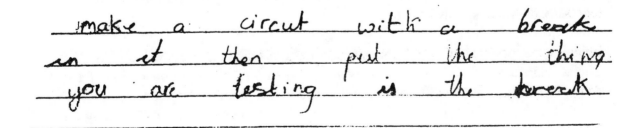

Fig 12: Sarah - Age 10.

However, many responses were incomplete or consistent with other ideas that children had about electricity and its flow. Fig 13 shows a response by a child who had earlier indicated that lighting a bulb required a single connection between the bulb and the wire.

Fig 13: Keri - Age 10

The vast majority of children though found this question difficult and failed to provide any answer.

The effect of more batteries

The final question used in the elicitation was an attempt to explore whether children held any intuitive notions of voltage or associated the number of batteries with the 'push' provided in an electric circuit. Children were presented with a drawing showing two batteries in series connected to a bulb and asked what they expected to see. The predominant response was that the bulb would light up e.g 'I would expect to see the light bulb light up.' and 'the bulb to light up because the batteries have electricity.' The response 'I would expect to see the bulb light up very bright' was comparatively rare.

However, there were some variations across the age range. Hardly any infants indicated that they expected the bulb to be brighter, most of them indicating that it would light up. With Lower Juniors though, the pattern was different with a majority of pupils predicting that the bulb would be brighter. This pattern may be anomalous as it was not sustained by the upper juniors where the majority of children predicted that the bulb would simply light up.

Evidently the presence of two batteries was not a significant factor for most pupils. This would suggest that many children were operating with a binary conception of a complete circuit which works and hence the bulb lights, or an incomplete circuit which does not work, and thus the bulb fails to light. The possibility that there may be gradations of functionality within a working circuit is not something of which they seemed to be aware or observed from this question. However, it must be noted that this item did not asking for a comparison with a circuit containing only one battery and further examination of children's understanding of voltage and batteries is necessary for a more definitive insight.

5. The Intervention Phase(May-June 88)

The previous chapter provides some insight into the range of ideas about electricity held by young children. Valuable as this is in providing a perspective of the conceptual framework children are using to make sense of their perceptions and observations, the aim of this research was to attempt to extend previous work in this field by devising a set of intervention activities which could be used by teachers to develop children's thinking towards the commonly accepted scientific understanding of electrical phenomena.

The rationale that underpinned the design of the intervention was that teaching and learning would begin with a phase in which children would be provided with an opportunity to articulate their own thinking and understanding about electricity. This was done by providing children with a range of activities that elicited their thinking through drawing, writing and discussion. A qualitative review of much of the data has been presented in Chapter 4. The data obtained from the elicitation was used informally to provide the teachers with a familiarity and understanding of their children's thinking about electricity. A set of structured activities was then provided which would allow children to explore electrical phenomena. All of these activities had a preliminary phase which required the child to hypothesise, predict or speculate about the behaviour of an electrical system using their existing knowledge. Further experiences then provided an opportunity, however limited, for the children to explore their thinking and experimentally test and evaluate their ideas against their observations in collaboration and discussion with their peers and their teacher. These experiences were designed to broaden their schematic knowledge, extend their vocabulary and, where appropriate, generate a conflict between their thinking and experience which would lead to a re-evaluation of their ideas.

The design of the activities for the intervention was influenced by three factors

(a)   A preliminary analysis of the data

(b)   A set of ideas defined by the 'scientific' understanding (Chapter 2 - 'Defining electricity') which would assist a child in developing an understanding of the scientific world view.

(c)   The teacher's contributions and ideas.

The preliminary analysis of the data showed that children held a wide range of ideas about the behaviour of electrical circuits. Many children used simple 'source-sink' models as a hypothesis about how electrical items should be connected to batteries. In addition, there was a lack consistency about their responses. Many children who could show successfully how to connect a bulb to a battery, could not repeat this when presented with an electric motor and battery. As a result of this data, it was considered that the specific knowledge of how to connect an electrical device to a power supply should be addressed by the intervention activities.

Secondly, the preliminary data indicated that many children, especially infants, lacked any clear understanding of which materials would conduct electricity. This uncertainty was apparent when children were shown a range of materials and asked to indicate whether they would conduct or not. A further indication came when children were connecting circuits and some incorporated connecting wires by touching the insulating plastic to the device rather than the bare wire exposed at the ends.

Thirdly, pupils had shown an awareness of a wide variety of objects, particularly domestic objects, which 'use' or 'work' by electricity. However, there was considerable uncertainty about the origin of electricity which came from wire, satellites, lightning as well as power stations.

These findings were then compared with and the framework of scientific ideas defined in Chapter 2 which the research hoped to assist in developing an understanding of by children. Intervention activities were then designed which were seen as being appropriate to children's existing level of knowledge and understanding and which essentially addressed the following areas.

    a.   The necessity for any circuit to have two connections to a device and an electrical power source.

    b.   Materials can be classified into those which conduct electricity and those which do not.

    c.   Electricity can be used for lighting, heating, moving and making magnets.

    d.   Electricity can be made in power stations using dynamos.

It was decided to directly address only these four and not the model of an electric current held by pupils. Children were encouraged to speculate and talk about the electric circuit using terms such as 'flow', 'continuous loops' or 'no break in the circuit' but no attempt was made in the intervention to examine systematically why two connections were needed. One of the basic difficulties faced in this area is that it is impossible to 'see electricity'. All models are inferences based on the effects of electricity and this level of understanding is an aspect which science education seeks to develop in the 11-16 science curriculum. The intervention activities were designed to assist in developing the foundations of an appropriate schematic knowledge which further experiences could build on. However, they were not provided to teachers as a proscribed teaching scheme but rather as a set of activities which teachers could use with children when appropriate to the child's starting point. Teachers were encouraged to always begin by providing an opportunity for the child to use their own ideas as a basis for investigation and prediction. The role of the teacher was to intervene with the suggested material when the child's ideas for exploration and investigation were not fruitful. Thus the role of the teacher was balanced between allowing the child total freedom to explore and providing specific didactic explanation. This is a difficult role which required finesse and experience. However, the starting point for exploration always lay in the children's thinking.

The teachers involved in the project were provided with an opportunity in an in-service session to trial the activities and suggest modifications. Some activities involved using apparatus which was unfamiliar e.g dynamos and large batteries and this session provided an opportunity to explore some of the practical difficulties. One of the main concerns articulated by the teachers was a concern about their own knowledge of this topic area. Many felt that this limited their ability to provide appropriate questions and guidance to children in their thinking. Ideally, with more time, it would have been valuable to run a session taking a constructivist approach to developing the teacher's own understanding as this session revealed a large area of uncertainty amongst the teachers.

Teachers were told that the activities provided were essentially a resource which they could use with children as appropriate. The intention was not to be prescriptive but to modify the activities as appropriate. Teachers were asked to encourage children to devise their own tests with the materials provided if the suggestion was suitable.

Activity 1: Making Connections.

In this activity, pupils were given a light bulb, electric motor, battery and connecting wires. Fahnstock[1] clips were provided to assist the making of connections to wires and the batteries. Children were asked to discuss and draw a picture showing how they would connect the battery to the bulb/motor to make it work. When this was completed they were encouraged to try out their ideas. When, and if they achieved success, they were invited to look at their original drawing and discuss their previous ideas in the light of the result they had just obtained with their peers and their teacher.

The intention of this exercise was that it would challenge the common idea held by many children that only one connection was necessary and force a re-evaluation of their thinking. The idea that two wires are necessary for an electrical device to work is a pre-requisite to developing ideas of current flow and conservation of current. The reason for using more than one device was to provide a wider range of experience so that children did not view the light bulb as a unique object. As well as a motor and a bulb, it had been intended to include a low-voltage electric buzzer for use by the children. However appropriate devices proved difficult to obtain.

Other activities included here were making an electromagnet and heating steel wool. In both activities, children were told a minimum amount of information necessary to

---

1. This is a product name for a type of crocodle clip sold in the U.K.

do the activity. Essentially, this was that an electromagnet could be made by passing electricity through a wire wrapped around a nail. Children were then asked to suggest a strategy for making an electromagnet and testing it.

Heating the steel wool was an opportunity for children to observe the heating effect of electricity through an enjoyable experiment. They were asked to devise a way of making electricity go through it and provided with a large battery, wires and connectors. Children were asked to note or draw the method they used which succeeded and to discuss why other methods may not have succeeded. It was hoped that both of these activities would help to develop the idea that two connections to a power source are necessary for any electrical device to function.

In practice, many teachers found that the apparatus often failed to make an effective electromagnet because of the high currents drawn from the battery to achieve an observable effect. Hence many children did not attempt this activity.

The activities in a second set were of a simpler observational nature. These involved examining bulbs and mains wires. Children were asked to draw what they would expect to see if they looked inside. They were then provided with specimens of each and allowed to cut open the wire and given a magnifying glass to look at the bulb and asked to sketch what they could see. It was hoped that the opportunity to see that a mains cable is not a single wire and that light bulbs have two wires going to the filament would help to support a model which saw devices requiring two connections to function.

A similar activity was devised with batteries. Children were provided with two batteries, a bulb and connectors and asked to show how they would make a circuit with two batteries in it. The opportunity was then provided to test such a circuit and observe its effect. Children were also provided with a range of batteries and asked to draw the batteries and note features common to all batteries. The batteries supplied varied in size and voltage. They were then asked to predict which would light the lamp most brightly and place them in an order. An opportunity was then provided to test the effect of using the different batteries with 4.5 V bulb which does not blow. This experiment was designed to challenge intuitive notions that the largest batteries are the strongest and to develop a tacit understanding that the brightness of the lamps followed the pattern of numbers with a capital letter 'V' after them.

b.   *Materials which conduct electricity*

An open-ended activity was designed for use with children. Children were given a bulb and holder, connectors and a battery and asked to work as a group and devise a way of testing objects to find out which ones let electricity pass through. Children were encouraged to test their ideas of how the bulb should be connected to function. Teachers were asked to assist pupils who had difficulty thinking of an appropriate mechanism for tackling the problem. Children were then asked to collect a range of common materials from their classroom and construct a table with their prediction for each maerial of whether it would let electricity pass and the answer found by testing it. The approaches to this activity reflected the range of styles that were used by teachers. Some allowed the children to work collaboratively in groups whilst some teachers preferred to work with the class as a whole, allowing them to predict and perform the experiment and acting as a central recorder of results. The activity itself provided rapid feedback as to the validity of their guesses.

An extension of this activity was to ask pupils to make a switch. Many pupils simply suggested breaking the circuit in some way and others made switches successfully from drawing pins and paper clips. The function of this activity was to develop a simple picture of a switch and reinforce the concept of a circuit which had been tackled previously. Children had to construct complete, working circuits before they could make switches. Unfortunately, there was insufficient time to explore whether children saw the position of the switch in the circuit as being important.

c.   *Where does electricity come from?*

This section of the intervention aimed to develop children's ideas about sources of electrical power or energy. Opportunities for practical work in this area are limited by the resources available to schools though hand operated dynamos were supplied to schools so that children could have an opportunity to explore generating electricity for themselves. It was decided that the main focus of the work here should be through collaborative work based on the use of secondary sources. Children were asked to discuss and write their ideas about the objects and places it was possible to get electricity from. A selection of books was provided and children told that they had to produce a poster with the heading 'Where electricity comes from.' The work was reliant on secondary sources but involved the children, through discussion, in the active construction of a report.

General issues

As with the similar phase of research investigating light there was a lack of any specific consistency between one classroom and a next. Teachers were entrusted to incorporate as many of the interventions as they could in the intervening period and there was inevitably variation in the time devoted to the topic and the extent of use of the intervention material provided. Some teachers chose to use the topic of electricity as a vehicle for doing the intervention materials and many other cross-curricular activities incorporating mathematics and English so the children's exposure to this concept area could be described as extensive rather than intensive.

Such variation is inevitable and and a reflection of normal classroom realities. Children were provided with an opportunity to consider their own thinking and test their own predictions and the data discussed in the previous chapter and the following chapter is a reflection of the empirical nature of the study. Consequently, the data can not be used to judge the validity of any one activity but merely provides one analysis of the potential developments in children's thinking from an exposure to a range of such experiences. The data were gathered from 6 classrooms with predominantly experienced teachers which places constraints on the reliability of the study. But this does not diminish the valildity of what was observed when such an approach is undertaken to the teaching of this topic.

6. The Effects of the Intervention

This section provides a full analysis of the data gathered during this study. A summary of the main findings is provided in section 8. Data presented here shows children's responses to questions about:

- a. Uses of Electricity
- b. Ideas about electricity
- c. Circuits and their connections
- d. Materials that conduct electricity and how to test for conduction
- e. The effect of more batteries on a circuit.

These data analysed here are those gathered in two phases, the elicitation phase prior to the intervention and a second elicitation phase after the intervention. In both phases, the elicitation work consisted of a large collection of activities which were designed to stimulate children to talk, write and draw their ideas about electricity and phenomena associated with electricity. Data from infant children were collected by interview due to the difficulty such children experienced in expressing themselves by other means.

In order to improve the reliability of the data, redundancy was built into some of the elicitation activities through the use of duplicated items that differed in their context so that the consistency of the responses provided by each individual child could be evaluated. The data analysis has incorporated this element.

The data presented are those obtained from children who were present on all three occasions i.e for the first elicitation, the intervention phase and the final elicitation. Consequently, substantially more data was collected than presented here. Full sets of data were collected from 107 children in total. Sample sizes for the different age groups varied considerably depending upon the availability of classes and children (n = 62 for upper juniors, n = 27 for infants, n= 18 for lower juniors). Whilst this spread was not ideal, difficulties were experienced in some schools due to staff mobility, timetable pressures and absences of children.

However, the data sample has been considered large enough to present a frequency analysis of many of the responses. Much of this was done using systemic networks (Bliss, Ogborn & Monk, 1983)[1]. Networks were evolved by comparison of the data with the suggested structure and critical evaluation of their effectiveness at representing the data. It is hoped that they present a considered attempt to provide an analysis of children's thinking at this age.

---

1.   Bliss J., Ogborn J. & Monk M. Qualitative Data Anlaysis. Croom Helm, 1983

Uses of Electricity

The elicitation activities had two specific items which produced responses about the uses or function of electricity. All children were asked 'What do we use electricity for?' and children older than 7 were asked to write three sentences with the word 'electricity' in. The former question tended to produce lists from children of typical items. The latter question was more open ended and responses such as 'electricity works lights' were considered a recognition by the child of a specified use.

In all, children mentioned 54 appliances that used electricity. These were cookers, lights, heaters or fires, television, irons, kettles, video recorders, fridges, radios, freezers, tape recorders, telephones, washing machines, hoovers, keyboards, hi-fi and stereos, toys, hairdryer, tumbledryer, microwaves, grills ,toasters, torches, computers, shavers, lawnmower, camera, batteries, motorboats, cars, machines, food processors, doorbells, plugs, switches, piano, buses, drill, aeroplanes, clocks, cement mixers, helicopter, machines, sewing machines, spinners, meters, buildings, tube (underground), typewriters, houses, motors, lightning, motor bikes, taxis, lorries, appliances.

	Pre			Post		
Items	*Infants*	*Lower Juniors*	*Upper Juniors*	*Infants*	*Lower Juniors*	*Upper Juniors*
	%	%	%	%	%	%
Cooker	22	50	39	48	50	56
Lights	56	61	82	56	41	66
Heaters/Fires	22	0	19	19	33	21
T.V	41	50	60	22	33	39
Irons	4	6	6	15	11	18
Kettles	0	0	5	4	0	10
Videos	0	0	6	7	6	15
Fridges	0	0	13	0	0	11
Radios	4	6	23	4	6	21
Telephones	0	11	13	4	22	11
Washing Machine	7	0	8	0	0	13
Microwave	0	17	11	0	0	5
Computers	4	11	10	0	0	3

Table 6.1. Percentage of children indicating items which used electricity.

The obvious feature of this list of items is that it reflects a preponderance of domestic items which shows that the main context for the development of a child's knowledge of electricity is the home. Only a few of these uses were mentioned by more than 10% of any age group and Table 6.1 shows which items these were.

It is clear from these figures that most children are able to specify a range of domestic items which require electricity to function. These responses were tested for significance to see whether there had been any change in the distribution or number as a consequence of the intervention. None were found to have any significance and this implies that children's ideas of the range of uses of electricity were not affected by the intervention. The intervention did not seek to extend children's knowledge of the range of uses of electrical energy so this result is not surprising.

Because of the imbalance of the samples, with the preponderance of data obtained from upper juniors, it is not meaningful to group the data into one total for the responses prior to the intervention and another for those post-intervention as such a method would be too weighted to the upper junior sample. Such a procedure has value in providing a view of the overall effect of the intervention.

Ideas About Electricity

The elicitation activities included a range of questions which asked children about the nature of electricity and its properties. Particular questions which elicited data were

> 'Write three sentences about electricity.'
> 'What is electricity like?'
> 'Where does electricity come from?'
> 'How fast does electricity go?'
> 'What do we use electricity for?'

The answers to these questions provided a large body of data reflecting of children's understanding of electricity. These data were summarised using network analysis, a method of categorising children's responses for the purpose of a quantitative analysis. The data are examined for clear categories of response produced by children and a network drawn up. Children's responses are then classified in terms of the categories of the network and counts made of the numbers of particular responses obtained. The network evolves through a process of successive approximation. Consistent failure of the network to provide a good representation of children's responses leads to its reformulation and another attempt to categorise the data. The final network evolved for representing children's ideas about electricity is shown in Table 6.2.

For example, analysis of children's responses identified three inclusive aspects of children's responses about electricity. These were ideas about the qualities of electricity, ideas about its origin and ideas about how it travels. These aspects are represented by an inclusive 'bra' which is shown in Fig 6.1. The use of the 'bra' symbol is a standard convention to indicate that these aspects are inclusive aspects of any individuals response.

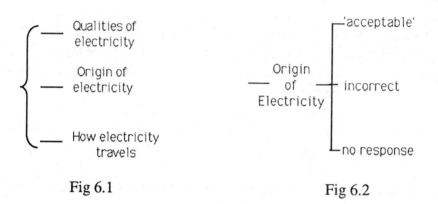

Fig 6.1   Fig 6.2

In the case of the origin of electricity, three distinguishable types of response emerged; those whose responses were 'acceptable' in a scientific sense, those whose responses were incorrect from a scientific perspective and those children which provided no response. For example, children who say that electricity comes from power stations have a defined perception of the source of electrical energy which is clearly separate from the child who says that it comes from lightning. The former response was coded as being 'acceptable' whilst the latter as being 'incorrect from a scientific purpose'. Since no child provided more than one type of response, these categories are mutually exclusive. The exclusive nature of the response is indicated by the use of a 'bar' as shown in Fig 6.2.

Fig 6.3

Finally some children provided more than one response about the 'qualities' that they associate with electricity. In order to represent the multiplicity of responses, this branch of the network has a recursive arrow (Fig 6.3). This indicates that the branch of the network is used more than once to represent a child's response. The bottom half of this 'bra' acts as a counter, indicating the number of times this branch of the network is entered for any one individual. Thus a child who said that 'electricity is dangerous', 'electricity is warm' and provides a specified use has indicated three qualities of electricity. Each one is marked separately on the respective branches (terminals) and a mark is made against 'three' to count the number of statements about the qualities of electricity for this child.

Networks are an instrument for data analysis and, as such are representations of the researchers perceptions of the children's responses, rather than those of the children themselves. For example, the division between 'qualities of electricity', 'origin of

		Infants (n=27)		Lower Juniors (n=18)		Upper Juniors (n=62)	
		Pre	Post	Pre	Post	Pre	Post
Qualities of Electricity	specified uses	24	24	17	15	61	58
	descriptive	9	19	16	11	38	29
	danger	13	8	4	5	20	29
	cost	-	-	-	-	3	-
	energy/warmth	14	9	1	7	13	13
	needed for living	-	2	1	2	6	20
	linked to gas	12	6	-	-	4	5
No mentioned	none	5	4	1	1	9	11
	one	7	9	14	2	29	17
	two	9	11	2	8	18	28
	three	4	2	1	6	6	5
	four	2	1	1	1	-	1
	five	-	-	-	-	-	-
	six	-	-	-	-	-	-
Origin of Electricity	'acceptable'	1	4	7	6	24	39
	incorrect — Scientific/technically associated	13	14	4	4	21	11
	incorrect — Non-scientific	7	7	3	3	12	9
	no response	6	2	4	5	5	3
How Electricity Travels	no response	11	6	1	3	14	10
	mode — other	10	8	9	4	12	12
	mode — wires	5	10	7	11	32	40
	mode — pipes	1	3	1	-	4	-
	speed — 'Acceptable'	10	22	16	13	52	57
	speed — incorrect	-	2	1	-	4	2
	speed — no response	17	3	1	5	6	3

Table 6.2: Network showing Children's ideas about Electricity

electricity' and 'how electricity travels' are classifications which are imposed from the researcher's perspective and which reflect the statements of the children.

The network shown in table 6.2 gives an overall view of children's ideas about electricity. Table 6.3 shows the median number of aspects or 'qualities' of electricity described by children.

	Infants		Lower Juniors		Upper Juniors	
	Pre	Post	Pre	Post	Pre	Post
Median No	3	3	2	2	2	3

Table 6.3: Median Number of statements about aspects of electricity by pupils

Predominantly these statements were either about specified uses or functions for electricity or of a more general descriptive nature. The following are examples of statements in the latter category:

'You cannot see electricity.'
'Electricity is like a blue streak of power.'
'Electricity is like water.'
'A good thing to use.'

The danger of electricity is a clear feature which was evident from the network, though in all cases it was only mentioned by a minority of children. A minority of pupils made statements linking electricity to gas such as 'electricity is like gas'. However, the predominant impression that emerges from an examination of the statements about the 'qualities' of electricity is the impression that electricity was seen by children as a vitalistic element, that is it is necessary for life, or an ingredient of machines, both of which are essential for human comfort and warmth. This would account for comparison with gas which is used for providing warmth and indicates that the children were intuitively recognising that both were sources of energy which are indispensable and both can produce warmth.

A statistical analysis of the network shows that there are only three significant changes for statements about the 'qualities' of electricity after the intervention. The number of infants who made descriptive statements about electricity rises from 9 to 19 of the pupils ($p < .01$); the number of lower juniors who made statements associating electricity with warmth and energy rises from 1 child to 7 ($p < 0.05$); and the number of upper juniors who made statements saying that electricity is 'needed for living' rose from 6 to 20 ($p < 0.01$). Given that there is no pattern to these changes and that in most instances, there was no change in children's statements, this does suggest that intervention had little effect on changing children's perceptions or models of electricity. This result was not be surprising since the data suggests that children's models of electricity are concrete in that the predominant aspects of electricity mentioned are everyday observable features e.g that it is used to make machines work; is dangerous and can be used for heating. These aspects would have predominantly been reinforced by the intervention activities.

Statements about the origin of electricity were categorised into 'acceptable' which was a broad category which included statements such as 'from the electricity house.'. A second category, 'incorrect', in which there were two categories of response, those that were technically associated e.g 'it comes from gas' and those that were clearly non-scientific e.g 'it comes from the sun' or 'it comes from lightning'. The final category was those children who were unable to give a response or gave an unintelligible response. The network shows that the majority of children are able to provide some response which, if not correct, has scientific associations and that the number of children providing such responses increases with age. However a statistical analysis of the network shows that a significant shift ($p<0.05$) has only occurred for upper juniors where the number of children providing an acceptable response has increased from 24 out of 62 to 39 out of 62. Whilst this is promising and indicative of a positive development, it shows that an understanding of where electricity comes from has not been developed for younger children. Given the previous evidence that children's thinking about the use of electricity is predominantly based in a domestic environment, and that approaches to developing any understanding of the origin of electricity are inevitably based on secondary sources, children's experience at this age has given them little opportunity to develop any understanding of the generation and production of electricity.

The final major feature of children's responses was their ideas about how electricity travels. The idea that electricity travels on wires clearly emerges as the predominant idea by the age of eleven. There were a few children who thought that it travelled in pipes either because they were confusing it with gas or more likely, given the urban environment in which the research was conducted, that they were correctly stating how they see electricity arriving. A large number of children produced other ideas about how 'electricity gets here'. Answers here varied from 'buying it in a shop', 'from the meter' to 'by satellite'. Such responses show clearly that for some children certain artefacts were associated with electricity but there was a lack of differentiation between one object and another in its purpose and function. This suggests that the schematic knowledge of the children is isolated and fragmented and lacks any model which enable distinctions to be made.

Children had a very clear impression that electricity travelled very fast and apart from the infants prior to the intervention, the majority of children appeared to know of this. Figures were often quoted in response e.g 100 miles per hour, 200 mph. A few children were questioned further about how they knew this and an explanation in terms of the rapid effect of a switch was often provided.

Statistical analysis shows that there was no significant change in the distribution of children's answers about the mode of travel as a consequence of the intervention. There was a significant change ($p<0.01$) in infants ideas about the speed at which electricity travelled. The number who provided an 'acceptable' answer increased from 10 out of 27 to 22 out of 27. The intervention did not directly address this idea but this result would indicate that it is one of the more perceptible features of the behaviour of electric circuits which infant children notice.

Overall, there are very few significant changes in children's ideas about the 'qualities' and behaviour of electricity as a consequence of the intervention. Since electricity is effectively imperceptible, all the concrete experiences of its behaviour and properties are of its effects and any understanding has to be inferred from these. The notion that it travels fast is easy to deduce from simple experiments with switches but an understanding of its origin, its mode of travel and use as a means of transferring energy are abstractions for many children which lack substantive evidence from their everyday lives.

c. Circuits and their connections

Much early education about electricity seeks to establish an understanding that a complete circuit is necessary for an electrical device to function. Consequently, the models held by children about the appropriate connections necessary to light a bulb or drive an electric motor were of particular interest. In the elicitation activities, three drawings were presented to children and the children asked to add to the drawing to show how they would get the bulb/motor to light. Many different responses were obtained which have been discussed in the Chapter 5.

The results obtained have been analysed by use of another network to provide a summary of children's understanding of the connections necessary to make electrical devices work (Table 6.4). The network shows the number of links and their associated arrangements together with the consistency of the response provided by children. It was hoped that this would provide some insight into the model being used by the child to generate a response.

The network shows that large numbers of children prior to any intervention use single connections between the battery and motor/lamp which reflects that the model being used by children is a simple source-sink model. This is shown more effectively in Table 6.5(a)-(c). The figures shown here are the percentage of the total responses of any one type, that is 33% of the infants responses prior to the intervention showed a single connection.

	Infants (n=27)		Lower Juniors (n=18)		Upper Juniors (n=62)	
	Pre	Post	Pre	Post	Pre	Post
no response	(37)	(16)	(9)	(7)	(30)	(11)
1 Connection	27	27	13	7	70	23
2 battery connections/1 device connection	7	7	4	7	22	28
2 battery connections/2 device connections	3	16	6	4	4	10
2 connections (correct)	7	15	22	29	60	114
no responses	9	-	-	-	-	-
1 response only	1	1	2	-	5	-
2 responses only	6	5	3	5	14	3
3 responses	1	-	-	1	21	19
2 responses	2	9	2	2	6	8
3 responses	6	7	10	5	12	27
2 inconsistencies in responses	2	5	1	5	4	5

Network structure:

Understanding of Connections
- No of links
  - no response
  - 1 Connection
  - 2 Connections (incorrect)
    - 2 battery connections/1 device connection
    - 2 battery connections/2 device connections
  - 2 connections (correct)
- Nature of response
  - Consistent response
    - no responses
    - 1 response only
    - 2 responses only
    - 3 responses
  - 1 inconsistency in responses
    - 2 responses
    - 3 responses
  - 2 inconsistencies in responses

Table 6.4: Network showing children's ideas about how to connect a circuit for an electrical device

	Infants	
	Pre %	Post %
No response	46	20
1 Connection	33	33
2 Connections		
2 battery, 1 device	9	9
2 battery, 2 device	4	20
2 Connections		
(Correctly indicated)	9	19

Table 6.5a: Nature of responses provided by infants showing how to connect an electrical device (%)

	Lower Juniors	
	Pre %	Post %
No response	17	13
1 Connection	24	13
2 Connections		
2 battery, 1 device	7	13
2 battery, 2 device	11	7
2 Connections		
(Correctly indicated)	41	54

Table 6.5b: Nature of responses provided by lower juniors showing how to connect an electrical device (%)

	Upper Juniors	
	Pre %	Post %
No response	16	7
1 Connection	37	12
2 Connections		
2 battery, 1 device	12	15
2 battery, 2 device	2	5
2 Connections		
(Correctly indicated)	32	61

Table 6.5c: Nature of responses provided by upper juniors showing how to connect an electrical device (%)

With the exception of infant children prior to the intervention, tables 6.5 (a-c) show that nearly all children used a more complex model with two connections to show how the device should be connected though only a minority were able to show how to attach the wires correctly. The implication is that it may not be helpful to start teaching electricity with bulbs where the two connecting points are not obvious. Children should be provided with an initial opportunity to investigate electrical devices to establish how many connecting points they do have.

Statistical analysis reveals that the changes in the responses of how to connect a circuit were highly significant for infants ($p<0.001$) and upper juniors ($p<0.001$) but the changes for lower juniors were not significant. This behaviour is somewhat anomalous but may be due to the small sample size used for lower juniors. Overall the results show that for all children, the changes were highly significant ($p<0.001$) though the sample was heavily weighted to upper junior children who showed a significant change in their responses. However this data shows that the provision of practical experiences with electrical circuits is a valuable component in developing operational knowledge.

The other half of the network was an attempt to examine how consistent children's responses were. This would provide some insight into the strength of the ideas they were using and the effect of context. The results are summarised in Table 6.6.

	Infants		Lower Juniors		Upper Juniors	
	Pre %	Post %	Pre %	Post %	Pre %	Post %
Consistent	63	22	28	33	65	35
One Inconsistency	30	59	66	39	29	56
No Consistent Response	7	19	6	28	6	8

Table 6.6: Percentage of responses from children and their nature.

These results show that for infants and upper juniors the effect of the intervention has been to decrease the consistency of the responses provided. The data for lower juniors were inconclusive. A very small contribution to the count for consistent responses was those children who provided only one response[1]. Such individuals cannot truly be said to have provided a consistent response. However their contribution would not change the overall pattern of results and it suggests that the effect of the intervention is to increase the range of responses and the context

---

1.   See terminal "1 response only" in Table 6.4

dependence of their answers. An examination of the data for the connections suggests that there was a decline in consistent responses which showed single connections which was accompanied by an increase in consistent responses which showed two correct connections. This effect was most marked with upper junior children.

One possible explanation of such results is that experiences provided for children by the intervention challenged their intuitive notions in specific contexts. Many children, realising the inadequacy of their thinking for *a specific example*, changed their response in this context to one which was more complex. This could be seen as a phase of confusion and was indicative that the child lacked sufficient schematic knowledge or ability to generalise from a limited range of experiences. In effect, the waters have been muddied but not changed and only those children who have developed an altered generalisable theory will show an improvement in their understanding with the use of a consistent response.

d. Materials that conduct electricity and how to test for conduction

One activity in the elicitation looked at the understanding children held of materials that conduct electricity. Children were shown a variety of materials and asked if they would let electricity pass through them. The main purpose of this activity was to see whether children were aware that there were a group of materials called 'metals' which conducted electricity. The responses provided by children are summarised in Table 6.7 (a-c).

Upper Juniors
(n=62)

	YES		NO		DON'T KNOW	
	Pre %	Post %	Pre %	Post %	Pre %	Post %
Wax	16	11	69	81	15	8
Cork	19	18	66	74	15	8
Comb	27	19	61	65	11	16
Scissors	63	85	23	6	15	8*
Foil	74	77	16	5	10	18
Paper Clip	76	90	11	5	13	5

Lower Juniors
(n=18)

	YES		NO		DON'T KNOW	
	Pre %	Post %	Pre %	Post %	Pre %	Post %
Wax	6	17	50	83	44	0**
Cork	11	17	72	72	17	11
Comb	17	0	56	89	28	11
Scissors	44	72	17	6	39	22
Foil	61	44	17	33	22	22
Paper Clip	56	72	22	22	22	6

Infants
(n=27)

	YES		NO		DON'T KNOW	
	Pre %	Post %	Pre %	Post %	Pre %	Post %
Wax	26	11	52	85	22	4*
Cork	37	26	37	67	26	7
Comb	22	26	59	74	19	0
Scissors	30	67	63	26	7	7*
Foil	41	85	41	7	18	7*
Paper Clip	48	67	41	26	11	7

** Changes which are significant at the level of p<0.01

* Changes which are significant at the level of p<.05

Table 6.7 a-c: % of children in each group and their responses to whether the specific materials would conduct.

The picture provided by the data is that many children already had a clear idea of which materials will conduct electricity and which materials will not, though this knowledge is more clearly defined with upper juniors. The intervention activities have produced some significant changes in understanding but since the pre-existing knowledge of many children was essentially correct, there was no substantial shift in their understanding. Those changes that did occur represent improvements in children's ability to differentiate non-conductors of electricity from conductors. The implication is that such an approach does not diminish any child's understanding and for some it has a positive effect.

Children were also asked how they would test a material to see if it would let electricity pass through. This was a difficult question for many children and responses obtained were often vague with essential elements such as a bulb or battery missing. Responses were categorised into four categories: no attempt, some attempt, nearly correct and correct. The distribution of responses is shown in Table 6.8.

	Infants		Lower Juniors		Upper Juniors	
	Pre %	Post %	Pre %	Post %	Pre %	Post %
No Attempt	63	33	77	33	39	19
Some Attempt	37	55	17	44	29	32
Nearly Correct	0	11	6	11	18	24
Correct	0	0	0	11	14	24

(Rounding errors have occurred in some percentages)

Changes for infants are significant at the 5% level and the data show that the trend in all cases was towards an increase in competency on this question. However only a small number of children were capable of correctly showing how the circuit should be constructed to test the material. This difficulty implies that their notion of a circuit may be specific to certain contexts and not easily generalised to unfamiliar situations. Alternatively it is possible that children found it difficult to produce a drawing which represents their thoughts rather than the weakness of their responses being a conceptual problem.

The effect of more batteries on a circuit.

This item was used to explore whether young children held a model of batteries that included at least an intuitive recognition of voltage. A bulb was shown to children connected to two batteries in series. It was hoped that children who had an intuitive notion that more batteries would drive a higher current because they had a higher voltage would have indicated this fact in their comments. The intervention had provided an opportunity for children to explore connecting circuits with more than one battery if they wished but this was not a specific activity that was recommended to teachers. Results are shown in table 6.9.

	Infants (n=27)		Lower Juniors (n =18)		Upper Juniors (n=62)	
	Pre %	Post %	Pre %	Post %	Pre %	Post %
No Attempt	15	7	22	6	10	11
'Lights up'	59	74	5	17	50	45
'Be Brighter'	4	4	55	72	26	35
Other	22	15	17	6	15	8

Table 6.9: % of children by age groupings and their responses indicating the effect of more batteries on the brightness of a bulb.

The predominant response for infants and upper juniors was that the light will light up though there is a sizeable minority of upper juniors who indicate that it will be brighter. Rather strangely, the majority of lower juniors recognise that the bulb will be brighter which is inconsistent with the other two groups. A possible explanation for this anomaly lies in the small size of the lower junior sample (n=18). None of the changes were significant and an examination of the figures shows that the intervention has done little to change children's knowledge of the effects of more batteries. This was a difficult area of knowledge to explore and in part the failure to produce any significant result may be due to inherent weakness of the item to place an emphasis on the presence of more batteries as opposed to the complete circuit.

7. Changes in Individual Children

The analysis in Chapter 6 provides an overall summary of the whole cohort but fails to provide any insight into the changes occurring for individual children. This chapter provides a view of some of the shifts in thinking that occurred for individual children which complements the description in terms of the networks.

The method is based on taking those items for which clear responses and categories of data are available and charting the changes that have occurred for each individual. This was done with the children's answers to items asking how connections would be made to bulbs and motors to make them function. The groupings used have been those of the network i.e:- a) no response to the item; b) one connection shown between battery and lamp; c) two connections shown with two connections to the battery and one to the device; d) two connections shown with two on the battery and two on the device but not a correct answer; e) two connections shown correctly. Data for changes in children's representations for upper Juniors are shown in Fig 7.1. The data are taken from the three items in each elicitation which asked children to show how they would connect the components so that they worked.

In the figure, the groupings of children's understanding are enclosed in circles. The arrows show counts for the number of children who have changed their response between the elicitation activities for that particular item whilst the number in the boxes, within the circles, shows the counts for the number of children who did not change their response.

The figures can be summarised into three groupings; (i) those which showed no change; (ii) those which show a change to a view which is indicative of progression-that is they changed from either no response to one connection for the bulb/motor or one connection to two connections though not necessarily scientifically correct; (iii) those which showed a less sophisticated representation. The charts show clearly the fluid nature of children's responses which not only changed from one context to another, but also from one period to another. The evidence is that children's responses can regress as well as progress.

Data for lower juniors and infants were analysed in a similar manner and the figures are summarised in Table 7.1. Children who had moved from a response which shows 'one connection' to 'two connections - correct' or, from 'two battery connections -1 device connection' to 'Two battery connections-2 device connections' were assumed to be showing a response which showed an understanding closer to the scientific model. Such responses were judged to show evidence of an awareness of greater complexity showing an awareness of the necessity of two connections which must be made to different points on the battery and the device.

Fig 7.1. Schematic chart showing changes in children's responses from pre- to post elicitation on how to connect electrical items to a battery.

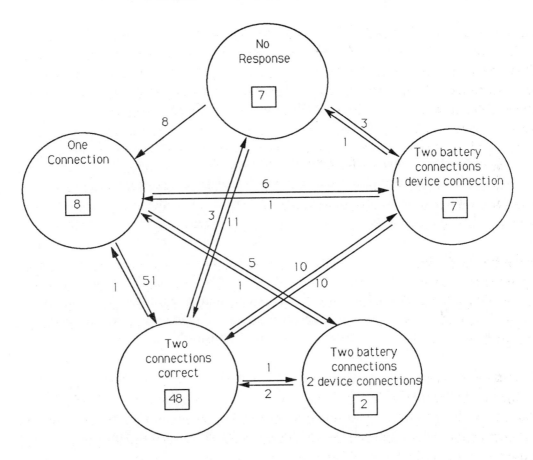

The most notable feature of these results is that the predominant trend was to more children providing a response in terms of two connections with relatively few children regressing. Those children which consistently hold the scientific view (or any other conception) are in a minority. The figures indicate that the predominant effect of the intervention was positive (Table 7.1).

	No Change Scientific Model	No Change	Change to a response closer to Scientific Model	Change to a response further from Scientific Model
UPPER JUNIORS	48	24	96	18
LOWER JUNIORS	22	9	16	7
INFANTS	2	27	43	9

Table 7.1. Summary of individual responses showing changes between elicitations to responses on connecting circuits.

A chi-squared test shows that there is a significant difference ($p<0.05$) for these responses as a whole between age groupings. Since the sample for the lower juniors was small, the significance was tested by collapsing the lower juniors with the infants. Most of the significance can be explained in terms of a larger number of upper juniors changing their thinking and a larger number already having a stable scientific conception.

The same method was used to look at individual changes exhibited by children in their understanding of the origins of electricity, the danger associated with electricity, the model of travel and the consistency of the responses that they provided.

The Origins of Electricity

The four schematic groupings used for this analysis are shown in Fig 7.2.

Acceptable	Not Scientific
Scientific/Technical	No Response

Fig 7.2: Schematic Categories used for analysing children's responses about the Origins of Electricity

The data for individual changes in children's responses to the origin of electricity are shown in Table 7.2. The response elicited from a child was assigned to one of the four categories above and then the process repeated for their response after the elicitation. From this it was then determined which of the categories shown in table 7.2 most appropriately described the change in their responses between the pre- and post - elicitation.

	No Change Response close to accepted idea	No Change	Change to a response closer to a scientific understanding	Change to a a response further from the scientific understanding
UPPER JUNIORS*	34	19	35	11
LOWER JUNIORS	28	28	16	28
INFANTS	4	48	37	11

Table 7.2: *Percentage of individual responses showing changes between elicitations to a question asking where electricity came from.*

* For tables 7.2 to 7.5 n=186 (Upper Juniors), n=54 (Lower Juniors) and n=81 (infants)

The figures indicate that, for the majority of children, the intervention has had no clear effect on the response they provided to this item. Only a minority of children provide a response which could be said to be 'acceptable' e.g that they indicate that electricity comes from 'power houses'. Not surprisingly, very few infants showed any knowledge of the origins of electricity. However, 30% of them did consistently provide a response which was scientifically or technically associated. A statistical analysis shows no significant differences between the changes from one age grouping to another. This suggests that children's ideas are relatively consistent and the intervention has had little effect in promoting change.

The Dangers of Electricity

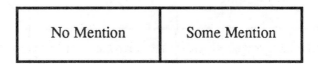

No Mention	Some Mention

Fig 7.3: Schematic Categories used for analysing children's responses about the Danger of Electricity

This aspect of electricity was not specifically addressed but was a prominent feature of the data. The data for the changes in individual children's expression of the idea that electricity was dangerous are shown in table 7.3. Again children's pre and post responses were coded to show whether they mentioned danger or not. The change in response was then assigned to one of the four categories shown in table 7.3.

	No Change Danger Mentioned	No Change Danger Not Mentioned	Change to a mention of danger	Change to No mention of danger
UPPER JUNIORS	19	41	27	13
LOWER JUNIORS	17	66	11	6
INFANTS	11	33	19	37

Table 7.3: *Percentage of individual responses showing changes between elicitations in mentions of the danger of electricity*

These figures show that the majority of children made no mention of the danger associated with electricity and only a minority consistently mentioned this aspect on both occasions. The intervention had little effect on their association of danger with electricity apart from infant children where a large minority moved to a position where danger was not mentioned as a quality of electricity. This change is the major contribution to the significance of the changes ($p < 0.05$). The possible implication, is that the opportunity to explore electrical components and circuits in a context where there was no danger associated with any of the items, diminished early associations between electricity and danger for very young children- an example of the association between ignorance and fear.

How Electricity Travels

No Response	Other
Pipes	Wires

Fig 7.4: Schematic Categories used for analysing children's responses about the How Electricity travels.

Children were asked in both the pre- and post elicitation how electricity got here and the data presented in Table 7.4 show how their individual responses to this item changed. Fig 7.4 shows the categories used for coding the responses and the headings of table 7.4 show the categories assigned to the changes in response.

	No Change Scientific Model e.g in or on the wires	No Change	Change to a response closer to the scientific view	Change to response further from the scientific view
UPPER JUNIORS	40	11	34	15
LOWER JUNIORS	33	25	28	22
INFANTS	4	31	48	15

Table 7.4: *Percentage of individual responses showing changes between elicitations to a question asking how electricity gets here.*

For the purpose of this analysis a change in response from 'no response' to one which indicated that electricity arrived in 'pipes', or a change from one which indicated that electricity arrived in 'pipes' to one which arrived 'on wires' was taken as evidence of an improved understanding by a child. There was a clear change in the number of children indicating that electricity arrives 'in' or 'on wires' from infants to lower juniors. Overall the intervention only affected a minority of children and statistical analysis shows that there is no significant difference between the distribution of responses across the age groupings. The change is positive for more children than it is negative, particularly for infants, but the data would indicate that the intervention has not been particularly successful in generating change in children's ideas apart from some limited success with infants. This would support the hypothesis expressed previously that an initial experience of electrical components and devices had a significant effect on developing children's knowledge of electrical phenomena.

Consistency of responses

Acceptable	Not Scientific
Scientific/Technical	No Response

Fig 7.5: Schematic Categories used for analysing the consistency of children's responses showing how to connect electrical circuits.

The final table, Table 7.5, presents data for the consistency of the responses provided by children when they were asked to show how they would make the appropriate connections to make bulbs and an electric motors work.

	No Change- Consistent Use of one Model	No Change in level of consistency	Change to a Consistent Response	Change to more consistency	Change to less	Response Rate Pre---Post % %
UPPER JUNIORS	27	18	11	3	40	83 93
LOWER JUNIORS	17	22	17	6	39	83 87
INFANTS	22	34	0	4	41	54 80

Table 7.5. Percentage of individual responses showing changes in the consistency of responses provided by children about how to connect electrical devices.

(Rounding errors account for summations indicating that the figures are not consistent with 100%)

The total number of responses provided by children pre- and post-elicitation were 44 and 65 (infants), 45 and 47 (lower juniors) and 156 and 175 (upper juniors) - which represents a 17% increase overall though the most substantive increase was in the number of infant children prepared to provide a response to items about electrical devices. Considered in conjunction with the results shown in the fifth colum (Change to less consistency), it is clear that one effect of the intervention, for a substantial minority of children, was to decrease the consistency of their response without any substantial increase in the number of responses. Only a few children showed a change to providing responses indicating the use of a consistent model. This evidence would support the idea that the response of many children was context dependent and that the wider range of experiences provided by the intervention has lead to the formation of models/ideas that were context specific rather than the formation of any model which has general characteristics. Statistical analysis shows that there is no significant differences in the increase in context dependence between the age groupings.

8. Summary

The following is a summary of the main findings described in section 6 and 7 and provides a resumé of the findings of this phase of the research. The data were obtained by an elicitation phase with children in the classroom from a mix of practical and verbal questions. This was known as the pre-intervention period. This was followed by an intervention phase when the children were encouraged to test their thinking and ideas through activities and investigations related to the topic. Following this period, another set of data was obtained from children using similar intervention activities which is referred to here as the post-intervention phase.

The main areas of note were found to be:

8.1 Children's understanding of electrical circuits

"When these students [American College students] were given a dry cell, a length of wire, and a flashlight bulb and were asked to get the bulb to light, most started by (1) holding one end of the wire to one terminal of the cell and holding the bottom of the bulb to one end of the wire, or by (2) connecting the wire across the terminals (i.e., shorting the cell) and holding the bulb to one terminal. They showed no sense of the functional two-endedness of either the cell or the bulb....It took 20 to 30 minutes for some member of the group to discover, by trial and error, a configuration that lighted the bulb....Seven-year-old children, incidentally, when given the same task go through exactly the same sequence at very much the same pace."

Arons, 1990[1]

Arons comments on his students' difficulties with the circuit concept and their prevalence from infant to higher education are an accurate reflection of the findings of this research and other work undertaken in this domain. The majority of children come to work on electricity with a concept of a circuit which can accurately be described as a 'source-sink' model. The effect of the intervention here has been to diminish, but not eliminate, the use of this model for lower juniors (7-9) and upper juniors (9-11). Clearly a positive indication that it is possible to alter children's conceptions in this area at this age. However, given the evidence for the tenacity of misconceptions in a minority of children and particularly for infant children, the question remains whether the effect is transient and teachers would be unwise to assume that one initial experience was sufficient to achieve change.

Another feature of the data was the influence of context on the response. Many

---

[1]   Arons, A. A guide to introductory physics teaching. John Wiley and Sons. New York.

Fig 8.1

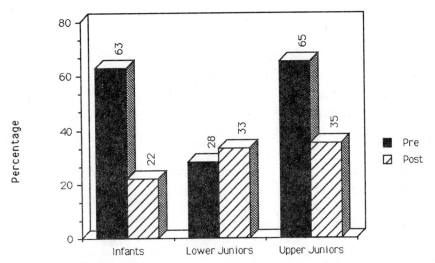

Chart showing % of children providing a consistent response to questions
about connecting electrical devices to batteries

children of all age groups both before and after, gave responses that indicated the use
of a context-dependent model. Fig 8.1 shows the extent to which consistent
responses were used by children before and after the intervention.

Given that, in the questions asked, there were no linguistic differences about how to
connect differing components to batteries so that they would function, the variation
suggests that the children lacked a generalisable concept of a circuit that would
enable the recognition of similarity. This is a disturbing feature of this work and
other work and would indicate the need for a broad range of experiences with a wide
range of differing components from which a general pattern could emerge.

This data also suggests that one consequence of experiences, such as those provided
by the intervention, is likely to be an initial increase in the lack of consistency. Only
the lower juniors showed a marginal increase in the stability of their responses. One
possible hypothesis is that the process of accommodation and equilibration which
results in the formation of a stable concept often produces a period of flux and
uncertainty which superficially can look as if the child's understanding has
diminished.

8.2 Properties and Uses of Electricity

The research examined the statements made by children about the properties and
qualities of electricity. Children provided 2 to 3 statements on average about these
aspects of electricity. The table beneath shows the predominant properties mentioned
by children and the percentages that made mention of them.

Property	Infants		Lower Juniors		Upper Juniors	
	Pre	Post	Pre	Post	Pre	Post
Specified Use of electricity	88	88	94	83	98	94
Descriptive statement	33	70	88	61	61	47
Danger of electricity	48	30	22	28	32	47
Electricity is needed for energy or warmth	52	33	6	39	21	21

Table 8.1: % of children who mention particular properties of electricity.

Nearly all children expressed the idea that electricity is used for a specific purpose be it lighting, heating or some other function. After that the other aspects that were substantially mentioned were the danger of electricity, the fact that electricity is essential for warmth or energy and descriptive statements about electricity e.g 'electricity is like magic'. How these varied across the groups is shown in Fig 8.2. These graphs reflect the dominance held by the function of electricity in children's knowledge. In the other categories, apart from a diminution in the descriptive statements across the age range, there was no clear pattern in the variation and it

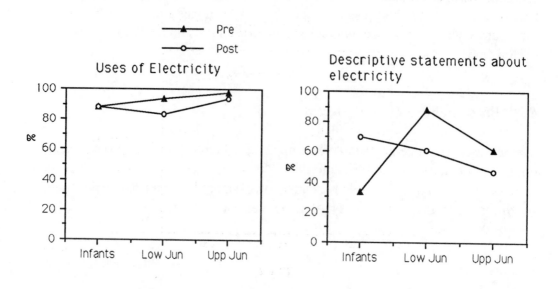

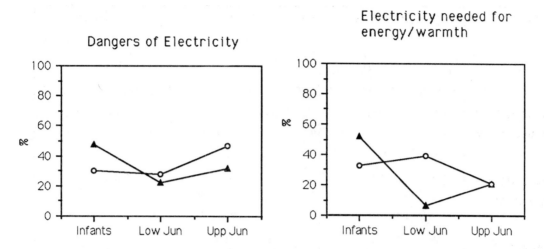

Fig 8.2. Charts showing the percentage of children mentioning common ideas about electricity

would seem that these aspects of electricity were those which dominated children's perceptions of the phenomena. The predominating impression was a world view of electricity based on simple observations and first hand experience.

That children of this age's knowledge of the properties of electricity was weak was typified in some of the statements that 'electricity is like gas', 'electricity is like magic' or 'comes via satellites'. Such statements possibly showed an attempt to describe electricity in terms of pre-existing constructs and illustrate the problem facing the child with a limited range of concepts and vocabulary. The evidence presented here and in Chapter 6 & 7 would indicate that the intervention did not significantly improve the models and understanding of electricity itself. However, this was not a primary aim of the intervention.

The one property of electricity that did seem to be well understood was the speed at which electricity travelled by all groups except infants prior to the intervention. Fig 8.3 shows that the intervention has significantly altered infants' perception of the speed of travel of electricity.

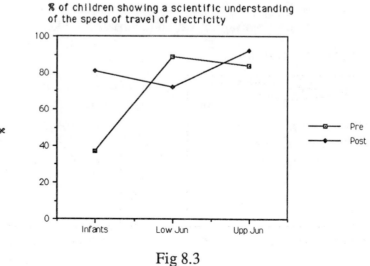

Fig 8.3

8.3   Knowledge of conductors and insulators

Another area of investigation was the pupils' understanding of which materials were conductors of electricity and which materials were insulators. 6 materials (3 conductors, 3 insulators) were used in all and the children asked if it the material would let electricity pass, would not let electricity pass or if they did not know. The data have been aggregated into conductors and non-conductors in Table 8.2 to show the responses obtained and their variation.

	YES will conduct		NO will not conduct		DON'T KNOW	
	Pre %	Post %	Pre %	Post %	Pre %	Post %
Conductors						
Infants	40	73	48	20	7	7
Lower Juniors	54	63	19	20	28	15
Upper Juniors	71	84	18	5	13	10
Insulators						
Infants	28	22	51	76	22	4
Lower Juniors	11	11	59	81	30	6
Upper Juniors	21	16	65	73	14	11

Table 8.2: Percentage of each type of response given by infants, lower juniors
and upper juniors about the ability of materials to conduct
(Data have been aggregated for 3 materials of each type)

The data show that the major changes in understanding of conductors and insulators occurred for infants and for lower juniors comprehension of insulators. On the whole upper juniors seemed to be cognisant of which materials are likely to conduct/not conduct electricity.

What all children found difficult was the application of the circuit concept to provide an illustration of how they could test materials for conduction/non-conduction. Only 24% of the upper juniors were capable of indicating an attempt that could be considered nearly scientifically correct. This leads to the conclusion that the application of circuit concepts in unfamiliar contexts is problematic for most children of this age.

8.4   The effect of more batteries - voltage

One item attempted to explore whether children had any understanding of the effect of adding more batteries to a circuit and hence by implication, voltage. Whilst the item may have failed to draw children's attention sufficiently to the feature of interest. The numbers of children who indicated that the lamp would light in the circuit would light more brightly is shown in Fig 8.4.

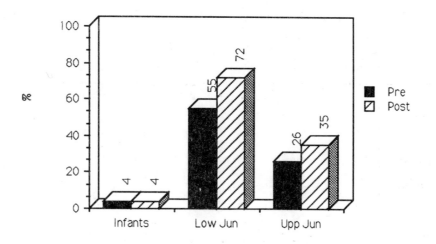

% of Children indicating that lamp would light more brightly

Fig 8.4

The data for the response shown by lower juniors is somewhat anomalous and must be considered in a context where the sample was small (n=18) and some doubt about the validity of the item. However, they do indicate that for the two older groups, the intervention had some positive effect though this was not significant.

Appendices

Appendix 1

ILEA

Inspector for Science Education	John Wray	

Schools	Headteacher	Teachers
Ashmole Primary	Ms P. Turnbull	Mrs M. Hutchinson
Benthal Junior School	Ms J. Auber	Ms J. Townson
Henry Fawcett Junior	Mr F.S.Curle	Ms R. Newlove Mrs M. Robinson
Johanna Primary	Mr J.W.Hines	Mrs D. Carter Ms R. Hines
Vauxhall Primary	Miss G.Brunt	Ms D. Gordon Mr V. Hayes
Walnut Tree Walk	Mrs V. Phahle	Mrs Wai-choo Tsang

Appendix 2:.
Questions used in initial phase of research.

The following questions were used in the first phase of the research, the pilot study, to explore children's understanding of the topic and to evaluate which questions were of particular value in eliciting responses from children.

1. Write three sentences about electricity.
2. Can you draw or name as many objects as possible which use electricity?
3. Where does electricity come from?
4. How is electricity made?
5. How does it get here?
6. What do we use electricity for?
7. What is the difference between electricity from plugs and electricity from batteries?
8. How does the switch on the wall work?
   Can you explain what happens when you turn it on?
9. How fast does electricity travel?

Activities

The following activities were tried with the children as a means of exploring their thinking about electrical phenomena.

Bulb, batteries and wires	Apparatus needed: battery, bulb, bulbholder and wire.
	How would you get the bulb to light?
	(This was tried with a single MES bulb on its own and a bulb in a bulbholder with two clearly visible 4 mm connectors)
Conductors	Apparatus needed: Pieces of wire (bare and insulated), plastic, aluminium, copper, wood, polystyrene, candle and string.
	Which of the above does electricity travel through?
	How could you test it?
Bulb in a circuit.	Show bulb lit by battery. Three batteries required-two same voltage and size, one same voltage, larger size.

What difference will a bigger battery make (ensure that voltage of battery is the same and marked clearly on it)?

What difference will two batteries make?

(Allow children an opportunity to test their thinking.

Static Electricity

Apparatus needed: Comb, balloon and pieces of paper.

Rub balloon and stick to wall.

Rub comb and pick up pieces of paper.

Does this have anything to do with electricity?

Explore any responses that state 'yes".

Appendix 3: Elicitation Activities

The following questions were used as the basis for the elicitation activities with children.

1. Where does electricity come from?

2. What do we use electricity for?

3. The drawing beneath shows a battery and a bulb. How would you get the bulb to light up?

4. How could you make the bulb light up using only a battery and one wire? Use the space below to do a drawing of your answer.

5. Write three sentences about electricity.
   (Tell me three things about electricity. (infants))

6. What is electricity like?

7. Will the following things let electricity pass through?

	YES	NO	DON'T KNOW
Wax			
Plastic Comb			
Cork			
Scissors			
Aluminium Foil			
Paper Clip			

8. This drawing shows a battery and a motor. How would you get the motor to work?

9. How fast does electricity go?

10. How does a switch work?

11. How would you test if a comb would let electricity pass through?

12. In the drawing, a bulb is wired to two batteries. What would you expect to see?

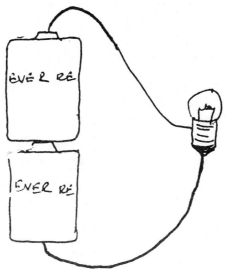

13. How does electricity get here ?

14. What is the difference between electricity from batteries and electricity from plugs?

Appendix 4: Intervention Activities

The following are a summary of the notes provided to teachers about the intervention activities to be used. Teachers were provided with an opportunity to try all the interventions at an in-service meeting. The importance of providing a stage for each activity in which children could discuss the task and consider their own thinking was stressed. Teachers were asked to encourage children to generate their own investigations to explore their understanding of electricity. These activities were provided as a support for teachers to use with children when judged appropriate.

Notes provided to teachers

The following notes are a guide to the main work that we would like you to do with your primary children on electricity in the next month. The aim of this work is to

a)   Develop an understanding in children that <u>two connections</u> are needed to make an electrical device work.

There are two subsidiary aims

b)   To introduce the notion that there is a complete path from the battery to the device and back again to the battery which is called a circuit.

c)   To develop the idea that there are possibly certain features which are commonly used to describe electrical supplies such as voltage and +(plus) and -(minus).

It is important in this work that the children *have an opportunity to test their own ideas out* as to how the electrical devices work. Whilst we see your role as providing guidance and assistance and suggesting possible solutions when they are stuck. Please give the children an opportunity to test whether their own ideas work before intervening and offering alternative solutions.

The following is a description of the suggested activities and an explanation of any of the difficulties that you may counter. Please try as many activities as you can. At the back are sheets that you may wish to use with the children to guide them through the activity.

Activity 1: Making Connections

This activity is designed to provide children with an opportunity to look at a wide range of electrical devices and see if they can get them to work. Each device requires two connections from the battery to the device to get it to work and this is the point that we hope children will observe. However, please do not force it but provide them with a wide range of experience so that they can develop this understanding themselves.

a. Lighting a bulb

Apparatus needed: Battery, bulb, wire, connectors

Pupil Activity: Before giving the children the apparatus, ask them to discuss how they think they will get the bulb and to do a drawing showing their ideas. *Please keep any such drawings with their names on if you can.*

Now give them the apparatus and let them try. Ask them to do a drawing to show how they did it. Give them some help if they really get stuck. Ask them how many connections were needed to make it work.

b. Making an electric motor work

Apparatus needed: Battery, buzzer, thick copper wire, connectors.

The instructions for this are exactly similar to those for lighting the bulb. The motor works with the battery connected either way. However, they should be able to spot that the motor goes the other way round when the battery is reversed. This may possibly lead to the idea that the electricity has a direction. When it goes through one way, it makes the motor go one way, when it is reversed, the electricity goes the other way round which makes the motor go the other way round.

Again, please keep any drawings that they do.

c. Making a magnet with electricity

Apparatus needed: **Large Battery**, nail, insulated wire connectors. Small needle or something which will be attracted by a magnet.

For this activity, the children will need the large battery. This is because to make an effective electromagnet, a battery which is capable of driving a higher electric current is needed. There is still nothing dangerous about it as the voltage is only 9V and you need to get to about 80V before you can begin to get a shock.

Ask children to discuss in small groups how they think they would do make a magnet with electricity and ask the children to do a drawing first which shows and then let them have a go. If they do not succeed, then please show them how to do it by wrapping a wire round the nail. The more turns the better and they should be using a piece of wire about 1 metre long as this will limit the current. The wire should not be left connected for too long as it will get hot and they can burn themselves.

Again they should be able to tell you how many connections they had to make in order to get it to work. Please get them to do a drawing showing how it worked.

d. Making things hot with electricity

Apparatus needed: **Large battery,** two wires, steel wool.

The large battery is needed for this activity as well. The children should be able to suggest how many connections they will have to make to the steel wool to pass electricity through it. Again ask them to do a drawing showing how they think they could use electricity to make the steel wool hot and then let them try it on the apparatus.

The correct solution is shown below. The wires merely need to be touched to the steel wool which should then get very hot and burn. The amount of heat generated is very small so there is no danger of anyone burning themselves.

Fig A4.1 Diagram showing how battery should be connected to steel wool.

Again, please get them to consider the question of how many connections are needed and do a drawing of how they succeeded.

Activity 2: Investigating wires, lights and batteries.

The purpose of this activity is to get the children to look more carefully at a variety of electrical components to develop a broader knowledge about electrical components and reinforce the idea that electrical devices need more than one wire to make them work.

a. Investigating wires

Apparatus needed: Selection of mains wires.
                      Scissors.

The children should be invited to speculate what the inside of the wire looks like and do a drawing of it. Then let them have the wire which they can cut open. Ask them to do a drawing of it.

Is it what they expect?
Why do they think there is more than one wire?

(Please remind children that on no account should they do this with a real wire. They risk killing themselves!)

b. Investigating bulbs

Apparatus needed: A large clear bulb or small torch bulb
A magnifying glass

Ask the children if they have ever looked inside the bulb.
What do they think it would look like?
See if they will do a drawing of what they think it looks like.

Now give them the apparatus and ask them to draw what they see. Is it what they expect? How many wires into the bulb are there? Is this what they expect?

Get them to look at the top of mains bulbs if you are using those. What is written on the top? If you are using torch bulbs, what is written on the metal casing where it joins the glass?

c. Investigating batteries.

Apparatus needed: A range of batteries of different sizes,
bulb
wire

Ask the children to look at the batteries.
What do the batteries have in common written on them?
Get them to do a drawing of each battery and write the common features under each one.

Now let them try lighting the bulb with the batteries. You will need a 4.5 V bulb supplied by us for this as these do not blow even if you use a 9V battery.

Is there any pattern between the brightness of the bulb and anything that is written on the batteries(The connection is that the higher the voltage, the brighter the bulb)

Let them see if they can use two batteries to light the bulb.
What is the effect of two batteries?

Electricity: Where does it come from?

The following are suggested activities to be used in the intervention to increase pupil's understanding of where electricity comes from and how it is made.
Activity 1

Ask pupils to find out where electricity comes from and how it is made. Start by asking them to discuss their ideas in a small group and present them to you on a piece of paper. Then ask them to find out what the answer is. They can ask at home, use books at school, home and the library.

Please can you give them some time to come back with the information which could be written.

Posters could be produced on how electricity is made and where it comes from.

Activity 2

The materials include a hand operated dynamo. Turning the end of the dynamo rapidly will produce sufficient current to light the bulb very briefly. A more sustained output can be provided by running it along the bench.

Children can be given the following questions to discuss.

When does the bulb light up?

Why does the bulb light up?

How long does it take the bulb to light up after turning on the dynamo?

Where are the two connections? One of the connections is very obvious and breaking this means that the bulb will not light. The other connection via the metal body of the dynamo is not self-evident and can the children show that there are really two connections by breaking the second one. This would mean undoing the bolt which may get lost unless looked after!

What happens if you undo the bulb? Is it easier or more difficult to turn. It should be more difficult but only just and you do have to know this to really be sure. However see if children can spot this. What it shows is that you have to work to produce electricity.

Useful Reference books for children

1.   Visual Science. Electricity Alan Cooper

Macdonald Educational   0 356 07113 6

2.   Let's Do Science: Magnets and Electromagnets   Malcolm Dixon.

Edward Arnold   7131 09068

3.   Science Exploration: Magnetism and Electricity

Ken Hutchinson.   Evans   237 293250

4.   My favourite Science Encyclopedia.

Hamlyn   0600 388 61 1

Conductors and Insulators

The aim of this exercise is to provide children some experience that some materials will let electricity pass through while others will not.

a. Testing for materials that let electricity pass through.

Apparatus needed: Bulb, batteries, wires and clips
     Variety of different materials including some metals.

Provide children with a selection of materials and tell them that you want them to find out if electricity will go through the material. Ask them to start by discussing how they will use the equipment you have to test it and to discuss with each other whether they think electricity will go through. Ask them to record their answers.

When they have done this, they should be allowed to test their thinking with the apparatus. They may need help to set up the correct circuit. Ask them to record their answers. They should be encouraged to try a wide variety of materials from around the classroom. When they have finished ask them to compare their answers with their guesses and discuss any they got wrong.

b. Making a switch.

This activity is essentially a technological project to see if they can apply knowledge about electricity to making a simple artefact.

Apparatus needed   Bulb, battery, wire, clips, drawing pins, wood block, paper clips.

Tell the children that the circuit they have made needs a switch so that they do not have to hold the wires together all the time. Provide them with the bulbs and batteries but also provide the other apparatus and challenge them to make a switch so that the light can be turned on and left on.

Encourage them to discuss how they think it should be done before trying. If and when they are successful, ask them to try other materials in the switch to see if that will work. Ask them to record any successful solution.